The Natural History of Ireland

T0073454

*Acknowledgements*

I wish to thank the following: John Barry, lecturer in Ancient Classics in UCC, for all the help and encouragement he has given me since I first undertook this project. His encyclopedic knowledge of the Classics and Latin authors had been invaluable. He also has a deep knowledge of Giraldus Cambrensis and Richard Stanihurst, as he has just translated the *De Rebus in Hibernia gestis* of Richard Stanihurst. In the *Zoilomastix* Don Phillip O'Sullivan attacks both Giraldus and Stanihurst. Without John's guidance I do not think I would have undertaken this translation. Professor Keith Sidwell formerly of the Department of Classics, UCC, who read the final proofs of the manuscript and made suggestions, and for his kindness in writing a foreword to the book. Kenneth Nicholls for his help with the names of towns, dioceses, rivers, mountains, etc., in chapter IX. Dr James Good DD who read the early proofs and offered suggestions. Paul McCarthy, a fellow student of mine when doing the BA, and who subsequently went on to do an MA in Latin, who read the early proofs and offered suggestions. Dr Jason Harris, lecturer in the Department of History, UCC, helped get the final document ready for publication. Dr Hiram Morgan of the Department of History, UCC, who read and commented on my introduction to the manuscript.

# The Natural History
# of Ireland
### included in Book One of the Zoilomastix
### of Don Philip O'Sullivan Beare

*Translated from the Latin by*
*Denis C. O'Sullivan BA, MB, FRCS, FRCSI*

CORK UNIVERSITY PRESS

*Dedicated to my dear wife Marie*
*for her encouragement and patience*

First published in 2009 by
Cork University Press
Boole Library
University College Cork
Cork T12 ND89
Ireland

This paperback published in 2020

Published with the support of the National University of Ireland

The author has asserted his moral rights in this work.

British Library Cataloguing in Publication Data

A CIP catalogue record for this book is available from the British Library.

ISBN Hardback – 978-1-85918-439-4
ISBN Paperback – 978-1-78205-396-5

Typesetting by Red Barn Publishing, Skeagh, Skibbereen, Co. Cork

Printed by Gutenberg Press, Malta

Cover image: A coat of arms for O'Sullivan Beare, detail from a copy portrait of Domhnall Cam O'Sullivan Beare, by P. Micopintoen 1859, donated to University College Cork in 1914 by F. J. Biggar and on display in the university. After an original portrait of 1613, artist unknown.

All paper-based materials used in the production of this book are grown in managed European forests under EU Regulations.

# Table of Contents

## BOOK I

**They discuss the Purgatory of St Patrick**

# Foreword

Denis O'Sullivan's translation of the first book of Philip O'Sullivan-Beare's *Zoilomastix* is a landmark. For many years, the Latin writings of the Irish of the early modern period have lain neglected, whether they appeared in printed texts (like Peter Lombard's *Commentarius* of 1631) or, like the *Zoilomastix*, in manuscript. At UCC, the *Centre for Neo-Latin Studies* has, since 1999, been undertaking the collection and study of this lost literature of Ireland. Between 1500 and 1750, when Latin was the medium of European intellectual discourse, more than 300 Irish writers produced more than 1000 printed works, and probably as many, if not more again, which, like *Zoilomastix*, never reached print (though this may not have stopped them circulating and having their own influence). I am pleased to be asked to write a foreword to the first of (we hope) a very large number of such translations (some with Latin text) which will appear in various places over the next few years.

*Zoilomastix* is a very difficult manuscript. It is hard to read and its many marginal annotations seem almost designed to baffle the editor. However, Denis has fought manfully to bring order to this chaos and has produced a clear version of a text that is at many places a revision and correction of O'Donnell's transcription and can be well defended as what O'Sullivan-Beare intended to be read. He has written a fairly literal rendition of the Latin, which serves to convey succinctly the information proffered by the text.

Denis has had a remarkable career. A career as a medical man was followed by the completion of a degree in Greek and Latin (with first-class honours) at UCC upon his retirement. The work before you is both the product of the skill acquired during that period and also a labour of love. For who better to present to the world a celebration of Ireland by O'Sullivan-Beare than an O'Sullivan, and at that an O'Sullivan from the same clan that produced one of the most important, and most neglected, Latin writers of the early modern period?

Keith Sidwell, Professor of Latin and Greek, University College Cork,
Founder of the Centre for Neo-Latin Studies and
the *Renaissance Latin Texts of Ireland* project.

# Introduction

## Ireland's Natural History

This unique contribution to the understanding of Ireland's natural history is part of the *Zoilomastix,* a manuscript held in the University of Uppsala in Sweden. It was originally written in Latin by an Irish nobleman in exile in Spain. The author did not set out to write about the flora and fauna of Ireland, rather it was part of a wider attempt by him to correct the blackening of Ireland's reputation by the medieval writer Giraldus Cambrensis in his *Topographia Hiberniae.* This Renaissance treatment of Ireland's natural history, which owes much to classical precedents, is here brought to the attention of a modern audience in a complete English translation for the first time. Its author was Philip O'Sullivan (1590–1636) from West Cork and it has its origins in the destruction of Gaelic Ireland in the aftermath of the battle of Kinsale (1602).

## The O'Sullivans of Beare

The O'Sullivans originally came from South Tipperary and had their castle at Knockgraffon.[1] In the twelfth century they were driven from their lands by the invading Normans. They eventually re-established themselves in West Munster in two branches – the O'Sullivan Beare settled in the Beara Penninsula and the O'Sullivan Mór in the Baronies of Dunkerrow and Iveragh in South Kerry.[2] Little is known of the O'Sullivan Beares until 1549 when Dermot, son of Donal, son of Donal, son of Dermot Balbh was killed in a gunpowder explosion.[3] He was the grandfather of our author, Don Philip O'Sullivan Beare and of his cousin, Donal Cam O'Sullivan Beare.

Dermot An Phudair, as he was called, left three sons by his wife, Sheela, daughter of McCarthy Reagh.[4] The three sons were Donal aged twelve,

---

1. *Annals of the Four Masters,* III, p. 94. Located two miles off the Cahir-Cashel road just south of New Inn.
2. Jeremiah King, *King's history of Kerry: history of the parishes in the county* (Wexford, 1912), pp. 265–279.
3. *Four Masters,* IV, p. 1133.
4. Carew Mss 635 at Lambeth Palace, London.

Owen aged nine, and the youngest Philip.[5] Dermot An Phudair also had a
son, Dermot (born 1532), by Sheela Fitzgerald to whom he was not married.
This Dermot, father of our author Don Philip,[6] was the eldest of the four sons
of Dermot An Phudair but did not succeed to the chieftainship, probably
because he was illegitimate. This would not have prevented him succeeding
to the title under the Brehon law but the O'Sullivans had for some time
followed the English system in relation to illegitimacy.

When Dermot An Phudair was killed in 1549, he was succeeded by his
eldest son, Donal. This Donal was killed in 1563 by a man called McGillicuddy.[7]
He was succeeded by his younger brother, Owen[8] who later became Sir Owen,
Lord of Beare and Bantry. This indicates that the Irish system of tanistry was
still in use among the O'Sullivan's at that time.[9] This led to a division of the
O'Sullivan Beare clan as Donal Cam, son of Owen's older brother Donal, laid
claim to the title under English law[10] which led to the division of the O'Sullivan
lands into Beare under Donal Cam and Bantry under Sir Owen.

Had Dermot, Don Philip's father, succeeded to the title as the eldest son,
although illegitimate, the divisions and hatred that developed subsequently
might have been avoided. In addition, this Dermot lived to be 100 years old
by 1632 when he died, so long-term continuity would have ensued.

### The Author: Don Philip O'Sullivan Beare

Don Philip O'Sullivan, the historian, was born in Dursey about 1590. Writing
in *The Compendium* in 1621 he states that he was a boy (*puer*)[11] when he left
Ireland in 1602. '*Puer*', according to the Lewis and Short Dictionary, is 'a
male child, a boy, a lad, young man (strictly until the 17th year but frequently
applied to those who are much older)'. It is generally thought that he was
probably born in 1590. Don Philip's age was given as twenty-four on the
16th September 1613 when he witnessed a document of recommendation
in Madrid. He signed another document in 1622 where his age was given as
thirty-three which indicates he was born in 1589. He signed a further letter
of recommendation in Madrid on the 20th December 1625, aged about
thirty-five years.[12]

---

5.    *Calendar of State Papers, Ireland*, III, 1586–88, p. 344.
6.    I refer to Philip O'Sullivan Beare, the Historian, as 'Don' Philip as this is how he is
      remembered by the people of Beara.
7.    *Four Masters*, V, p. 1597.
8.    *Cal. S.P. Ire*, III, 1586–88, p. 346.
9.    The system of succession by seniority rather than primogeniture. Tanistry was a major
      target of Tudor reforms.
10.   *Historiae Catholiae. Hiberniae Compendium.*, pp. 147–8.
11.   *Hist. Cath. Hib. Comp.*, Tom. III. Lib. VII. Cap. 1.
12.   Unpublished papers of Michelene Kearney Walsh from the O'Fiaich library, Armagh.

His father was Dermot O'Sullivan who was chieftain of Dursey, born in 1532. Dermot's mother's name was Síle Fitzgerald[13] and it would appear that her son, Dermot (Don Philip's father) was illegitimate, according to vol. 636 of the Carew Manuscripts at Lambeth Palace.

Don Philip was the son of Dermot of Dursey and Johanna McSwiney.[14] They produced seventeen children; fifteen boys and two girls. Don Philip, in a poem attached to the *Patriciana Decas* states[15] 'they say the dying swan sings a song. So I sing the sad death of my dear ones. My parents united by the ties of wedlock gave birth to seventeen children, dear tokens of love, but the dark day of death cut off thirteen in the flower of youth, when my country was still unharmed. My parents, exiles in Spain, beheld but four after the mournful fall of defeated Ireland.' Before going to Spain, Don Philip was educated at home by Donagh O'Croinin who was executed in Cork, 1601. In the *Zoilomastix*, Don Philip says he was hanged and eviscerated and his entrails thrown on a fire.[16] Don Philip was sent to Spain with Donal Cam's five-year-old son, Donal, under the care of Dermot O'Driscoll in February 1602 following the defeat at Kinsale. They arrived safely at Corunna and were hospitably received by Luis de Carillo, the Count of Caracena and Governor of Galicia, a great friend of the Irish.[17] Don Philip was educated at Compostella, of which he says: 'Compostella,[18] a teacher wondrous skilled in sacred literature educated me in her College. Patrick Synnott, cultivated in the Latin language, taught me the first elements of grammar. This man distinguished alike for his compositions in prose and verse, sudden death has recently carried off. Next, the venerable Vendana with exquisite skill unfolds to me, his pupil, the books of Physics. Marcilla, renowned for the religion of Peter, explained to me the hidden mysteries of the Divine Wisdom. Next, when engaged in the wars of the Catholic King by land, it was my lot also to encounter danger by sea.' Following his schooling he joined the Spanish Navy and received a commission from Philip III.

Of the four children remaining, following the deaths of his thirteen brothers, Helen was drowned on a return voyage to Ireland. He says[19] 'now

---

13.   Poem on family attached to Philip O'Sullivan, *Patritiana Decas*, (Madrid, 1629)
14.   *Ibid.*
15.   *Ibid.*
16.   *Zoilomastix*, f. 270v, Lib. III.
17.   *Calendar of State Papers, Spanish*, IV, 1602, p. 708. For a treatment of Carillo see Ciaran O'Scea, 'Caracena: champion of the Irish, hunter of the Moriscos' in Hiram Morgan (ed.) *The Battle of Kinsale* (Bray, 2004), ch.13.
18.   *Comp. Hist. Cath. Hib.*, Tom III. Lib VII. Cap. I; 1621 Lisbon ed. Poem attached to *Patritiana Decas*.
19.   Poem attached to *Patritiana Decas*.

my sister Helen, a second time married, is drowned, while she returns to her country. She was a woman of persuasive eloquence worthy of praise for her piety and modesty and endowed with an unspotted character.' His brother Donal was killed on 2nd July 1618 during a battle with Moorish pirates. Don Philip did not hear of his only surviving brother's death for about fifteen days because he was hiding in the French embassy following the death of Donal Cam, the Count of Birhaven. He describes how his brother was shot through the chest by a musket ball as he was leaping on to a Moorish ship. He says that his brother Donal was very handsome, very strong, and a good boxer, as well as that he had greatness of soul and was well educated with a good knowledge of Latin philosophy and logic. Thus was left only Philip and his sister Leonora and their parents. Philip adored his father, Dermot,[20] as a great hero and fighting man. He tells us that his father Dermot led men from Beara to the aid of Gerald Fitzgerald during the Desmond rebellion. He says[21] 'Gerald stormed and dismantled Youghal, a noble and very wealthy town, in the storming of which Dermot O'Sullivan, my father, captain of the foot of Beara with signal valour and against immense difficulties, scaled the walls by ladders, the besieged town in vain resisting.'[22]

> 'Shortly after this, Dermot accompanied by five retainers fell in with one of the Queen's magistrates accompanied by fourteen soldiers and a sharp fight took place. Finally, Dermot was struck down covered with many wounds. Two retainers were killed and many wounded. Nor was the fight a bloodless one for the enemy of whom no fewer fell with their leader before it was terminated by some men coming up from the nearest hamlets. Dermot and the other men were cured by most attentive nursing.'[23]

In a letter to Patrick Synnott he says, regarding his father Dermot, 'he was a hero most renowned in arms; he was most powerful both by land and sea, sometimes alone. The wide mouths of the Shannon often saw him victorious and were bewildered at the valiant exploits of the fearless chief. Borne in a single light galley, with shining sword, he repulsed fourteen armed vessels. He was content to overcome the enemy. He spurned wealth.'[24] He recounts how he built the boat which carried the O'Sullivans across the Shannon in their heroic retreat from Glengarriff in 1602. To his son he must have looked like the Ajax of Sophocles, a man of action and of few words, as few are

---

20.    *Ibid.*, Tom. 2, Lib. 4, Cap. 15, F01.98r.
21.    *Hist. Cath. Hib. Comp.*, Tom II. Lib. 4. Cap. XV.
22.    *Ibid.*
23.    *Ibid*, Tom. II. Lib. 4. Cap. 22; M.J. Byrne (trans.), *Ireland under Elizabeth, chapters towards a history of Ireland in the reign of Elizabeth being a portion of the history of Catholic Ireland by Philip O'Sullivan Beare (Dublin, 1903)*, p. 38
24.    Poem attached to *Patritiana Decas.*

recorded. Of his father, he says[25] 'When his life was prolonged to about a hundred years, he dies at last, his limbs being relaxed by old age; and the fortified city of the Galician Corunna contains his bones which were duly buried in the temple of the Seraphic Father.' Philip's mother did not long survive her husband. Philip, in anguish, says[26]

> 'Quam sum perculsus morte, Johanna! Tua!'
> *Oh! How I was affected by thy death Johanna!*

He laments in a poem about his family:[27]

> Funera post lachrymosa patrum fratrumque meorum
> Una mihi superset nunc Leonara soror . . .
> *After the sad deaths of my parents and siblings*
> *There is left to me now one sister Leonora . . .*

Don Philip himself appears to have remained unmarried all his life. He died in the second half of 1636.[28]

### The Death of O'Sullivan Beare

There is a lot said about the death of Donal Cam O'Sullivan Beare on 16th July 1618. The only contemporary account we have is that of his first cousin, Don Philip O'Sullivan Beare, our author. The latter was himself involved in a duel with John Bathe over some insults Bathe threw at Donal Cam in Philip's presence and which in the end resulted in the death of Donal Cam O'Sullivan Beare. I thought that I should give the original Latin and my translation from a copy of the original 1621 edition of the *Historiae Catholicae Iberniae Compendium* Tome 4, Liber 3 Caput IV, Folio 262V.

John Bathe was treated kindly by Donal Cam but an argument arose in Don Philip's presence over money that Donal Cam had given Bathe. These are Don Philip's words:

> Quorum beneficiorum Johannes immemor, eo
> impudentiae processit, ut levi primum controversia orta
> ob pecunias ab Osullevano mutuo datas, inde sit ausus
> tanti viri clarissimae nobilitati genus suum apud Ibernos
> & Anglos, a quibus oritur, minime sublime conferre. Quod
> aegre ferens Philippus OSullivani Patruelis, qui hanc

---

25.    *Ibid.*
26.    *Ibid.*
27.    *Ibid.*
28.    Information from Ciaran O'Shea, who is completing his doctorate on Irish exiles in seventeenth-century Spain at the European University Institute in Florence.

historiam scribit, cum Johanne ea de re expostulat. Unde Madriti iuxta regium monasterium Divi Dominici uterque alterum stricto gladio aggreditur. Incepto certamine Johannes ingente pavore percussus, & vocem efferens loco semper cedebat: & illum in facie Philipus caesa vulneravit interfecturus videbatur nisi eum Edmundus Omorra & Giraldus Macmoris ab Osullevano missi & duo equites Hispani protexissent. Philipum quoque apparitor deprehendisset. Cum multi undique confluxissent, inter caeteros O'Sullivanus advenit laeva manu Rosarium, & dextra chirothecas gerens. Quem Johannes conspicatus incautum, nihil timentem, & alio aspicientem subito accedens gladio inter turbam intento per laevum lacertum confodiendo & rursus guttur feriendo occidit. Philippus lictore frustra reluctante in domum Marchionis Senecciae Galliarum legati se abdidit. Johannes in carcerem conijcitur una cum consanguineo suo Francisco Batheo, qui rixa interfuit: sicut & Daniel ODriscol Philipi consanguineus.

*Unmindful of these gifts John acted with such shamelessness that, as soon as a trifling argument arose concerning moneys that O'Sullivan (i.e. Donal Cam) had given him as a loan, that thus he dared to compare his family, among the Irish and the English from whom he had his origins, to the outstanding nobility of such a great man (i.e. Donal Cam) in a derogatory fashion. This, Philip, a cousin of Donal Cam on his father's side, who wrote this history expostulated with John concerning this matter. Consequently, near the Royal Monastery of St Dominic in Madrid, they attacked each other with drawn swords. Once the fight started, John was trembling with great fear and, shouting out, he constantly gave way: and Philip wounded him with a cut in the face and it appeared that he was about to kill him, but Edmund O'Moore and Gerald MacMorris, and two Spanish noblemen, who were sent by O'Sullivan (i.e. Donal Cam), protected him (i.e. Bathe). An official of the law arested Philip. Then many people gathered from all sides, among others O'Sullivan arrived carrying rosary beads in his left hand and gloves in his right hand. When John saw him off his guard and not anticipating trouble and looking elsewhere,*

*John came up suddenly, through the crowd, with his sword stretched forward, he killed him (i.e. Donal Cam) by wounding him, on the fleshy part of the left upper arm and again by striking his throat. Philip, the official of the law in vain struggling with him, hid himself in the house of the Marquis Seneccia, the French Ambassador, John Bathe was thrown in jail along with his cousin, Francis Bathe, who was present at the fight, as also was Daniel O'Driscoll, Philip's cousin.*

This is the exact account of the death of Donal Cam O'Sullivan Beare, Lord of Beare to use his English title and Count of Birhaven to use his Spanish title. It is a dramatic account but very short on detail. Don Philip does not tell us why Bathe was so angry. The fact that he puts in the word 'mutuo', as a loan, in referring to the money may mean that Bathe felt he should have got it as a gift and that he may have complained that Donal Cam was a skinflint. This is pure speculation and we just do not know. How John Bathe escaped being tried and executed for Donal Cam's murder again we do not know. We know that he returned to Dublin, his home, and got a pension of £500 annually from king James I.[29] Recent research by Dr Hiram Morgan has discovered that the Spanish king passed by as Donal Cam was killed, and, on hearing the Irishmen bewailing their leader's death, ordered that Bathe be arrested and the death investigated. The result of the investigation has not been found. Bathe, writing in his own defence to the Count of Gondomar, implied that he was attacked by a group of Irishmen and was forced to defend himself.

In his account, Don Philip goes on to say that

*Donal Cam was buried the following day in the presence of a large gathering of Spanish noblemen and the funeral expenses were paid by Don Diego Brochero, a splendid knight and a councillor of the King.*Don Philip finishes this chapter by saying 'He was *fifty-seven years old when he died, he was clearly a pious and generous man especially to the poor and destitute. He was accustomed to attending two or three masses daily, pouring forth long prayers to God and to those in Heaven: frequently when his sins were forgiven, he received the sacred body of the Lord. His sudden and unfortunate death was not at all consistent with his life; even on the very day that he died he attended two masses, and when he was wounded he was absolved of his sins by a priest. In stature, he was tall and elegant, with a handsome face and venerable, at an age when people become grey.*

---

29.　Micheline Kerney Walsh, *'Destruction by peace': Hugh O'Neill after Kinsale* (Armagh, 1986), pp. 106–13.

It was already believed in Irish émigré circles that John Bathe was an English spy. On 12th April 1618 Florence Conroy, the Archbishop of Tuam, as he was about to leave Madrid for Flanders wrote to King Philip explaining to him why he thought John Bathe was a spy. The Archbishop's letter was discussed at the Council of State on 16th May 1618 when it was recommended that Bathe should no longer be trusted or admitted to court. No further action was taken in that regard when on 16th July 1618, following an argument and sword fight with Philip O'Sullivan, John Bathe stabbed Donal O'Sullivan Beare.

In a letter that Don Philip wrote to his cousin Dermot on the death of his father, Donal Cam, he states:[30]

> *To the most ilustrious Don Dermot O'Sullevan, Count of Dunboy*
> *Philip O'Sullivan*
> *Health Through God*
> He signed off at the end:        *Farewell Cadiz Ides of April 1619*

He sympathises with his cousin on the death of his father, Donal Cam. Regarding Bathe he points out to Dermot that great people in the past, such as great Alexander and Julius Caesar, have been killed by members of their own household. He then says 'therefore it is no way wonderful that your brave and illustrious father was murdered by that wretch, whom he assisted and furthered by acts of kindness, who sometimes had a plate at his table and who perhaps dipped his hand in the same dish with him.'

On the Ides of May 1619, he wrote to Fr Patrick Synnott:[31] 'The murder of the Count I took to heart more grievously than anyone could imagine nor indeed at the present moment am I less afflicted with more grief than on the very day on which he was slain. In a few days after the quarrel when having escaped from the hands of the officials of the law, I had taken refuge in the house of the French Ambassador, I was informed of the death of my brother Daniel and of Philip and of my other friends.'

His description of the death of Donal Cam is very skimpy and perhaps the Spanish archives have still to yield up greater detail to some research student who puts his or her mind to it.

While living in Spain he would have been regaled by his father Dermot, his mother Johanna, his cousin Donal Cam and all the other people in the O'Sullivan household about the battles fought, the defeat at Dunboy and their lands stolen, their property destroyed and famine inflicted on those left behind by the scorched earth policy of Mountjoy and Carew. This scorched earth policy had been first proposed by that gentle poet, Spenser, who while

---

30.    *Hist. Cath. Hib. Comp.,* Fragments attached at the end, p. 265V–266R.
31.    *Ibid.,* Fragmenta F.270 V.

living in Cork, having been driven from Kilcolman after the battle of the Yellow Ford in 1589 wrote 'great force must be the instrument but famine must be the means. For till Ireland be famished it cannot be subdued.'[32] The hope of returning to Ireland with the help of the Spanish was a hope that always survived in the hearts of the exiles. All this fuelled the flames of patriotism in Don Philip's breast and fostered a hatred of the English whom he referred to again and again as heretics. One must remember that in 1532 when Dermot, Don Philip's father, was born, official English Protestantism did not exist and Henry VIII was a devout Catholic until 1535 when he broke with Rome. Following his break with the Pope, the Four Masters, *sub anno* 1537, state 'A heresy and a new error sprung up in England, through pride, avarice and lust and through many strange sciences so that the men of England went into opposition to the Pope and to Rome. They destroyed the orders to whom worldly possessions were allowed, namely the monks, canons, nuns, brethren of the cross, and the four poor orders, i.e. the Orders of the Minors, Preachers, Carmelites and Augustinians, and the Lordships and livings of all these were taken up for the King. They broke down the monasteries and sold their roofs and bells.' By this action they destroyed the monastic schools in Ireland. The Four Masters continue, 'they afterwards burned the images, shrines and relics of the saints of Ireland and England; they likewise burned the celebrated image of the Blessed Virgin Mary at Trim.'[33] This was the start of the Tudor Conquest of Ireland which brought suffering, famine, death and abject poverty to the Irish people.

### Don Philip's Writings

1. Don Philip is best known for his *Historiae Catholicae Hiberniae Compendium*. Usually referred to as the *Compendium,* it is also known as 'O'Sullivan's Catholic History'. This was published in 1621 in Lisbon. It is Don Philip's most well-known publication. He traces the origins of the Irish back to the Milesians whom he says came from Spain, emphasising the connection between the Irish and Spanish in the hope that Philip IV of Spain might be inspired to assist the Irish in driving the English from Ireland. He also insists on the loyalty of the Irish to the Catholic Church, from the time of St Patrick onwards. He describes how bishops and priests suffered torture and death rather than accept the Protestant religion. He also recorded the events of the Nine Years War led by Hugh O'Neill and Hugh O'Donnell with the eventual defeat at Kinsale. He relates how his cousin Donal Cam had

---

32. *Cal. S.P. Ire., VII,* 1598 p. 433.
33. *Four Masters,* V, pp. 1445–7.

handed over his castles to the King of Spain but had to retake them lest they be handed over to the English, and he describes the massacre of his own people on Dursey Island during the siege of Dunboy, and the resulting epic march of Donal Cam and his people to O'Rourke in Breifne, which only thirty-five out of about 1000 people survived. The *Compendium,* based on eye-witness accounts and on ancient manuscripts, was passed for publication by the Inquisition.

2. The *Zoilomastix* written in manuscript form and now being translated. This was finished in the first half or so of 1626.

3. The *Patritiana Decas,* published in Madrid 1629, a history of St Patrick in ten books. There is a long poem about his family attached to it. It also contains a reply to Archbishop James Ussher who, when he read the *Compendium,* called Don Philip 'the most egregious liar of any in Christendom.' This reply is called the Archicornigeromastix, in which he castigates Archbishop Ussher.

4. A Life of St Mocuda, published by Bollandus in Antwerp, 1635.

5. A Life of St Ailbe was rejected from the *Acta* on the advice of Hugh Ward as he felt it contained errors.[34]

6. He started a work on astronomy but abandoned it.

7. Tenebriomastix, held in the Mediathèque François Mitterand (MS259/97) in Poitiers in manuscript form, has been translated by David Caulfield at University College Cork. This work is dated 1650, presumably the year it was copied for the Irish college at Poitiers. The Tenebriomastix is a refutation of the claims of the Scottish writer David Chambers' 'De Scotorum fortitudine, pietate et doctrina,' Lyon, 1631, in which he claims that Irish saints were actually Scottish. The Irish were referred to as 'Scoti' by early writers, and Chambers claims that the reason is that the Scottish colonised Ireland, rather than the other way round. Don Philip vigorously refutes Chambers' claims, along with those of earlier Scottish historians such as Thomas Dempster, Hector Boece, and John Major. These Scottish writings drew responses from several other eminent Irish authors of the seventeenth century, such as Stephen White, David Rothe, Thomas Messingham, John Lynch, Geoffrey Keating and Roderick O'Flaherty.

8. In Manuscript 580 Trinity College Library Dublin, there is a report entitled 'A BRIEF RELATION TO IRELAND' (written in capitals) 'And YE Diversity of Irish in the Same' (this added in a regular writing)

---

34. *Analecta Bollandiana,* Vol. 1, p. 47.

in the catalogue. This is attributed to Don Philip O'Sullivan Beare. The report is in Folios 95–98 of the manuscript. In the upper left hand corner of the first page is an entry in small writing in the handwriting of James Ussher, the Protestant Archbishop, which reads 'ensented to in Council of Spayne circum. 1618 by Florence the pretended Archbishop of Tuam and thought to be penned by Philip O'Sullivan Beare.

9.  In 1625 Don Philip made a presentation to Philip IV of Spain, showing how Ireland could be conquered (Paris, Affaires Etrangérès, Éspagne 254/142 ff228–233). He refers to himself as Lord of Piñalba in Ireland, and also to his father Dermot by the same title. The plan for the conquest was approved by his father, Donal Cam, and Don Cornelius O'Driscoll. Don Philip designed a plan for conquest beginning with Dursey Island, and goes on to describe the time and resources required. He requests that he be placed in charge of a local regiment to be raised in Ireland, pointing out that the English hate him because of the publication of his Compendium in which he revealed their wickedness and persuaded the Irish people to separate themselves from their tyrannical, heretical government. His plan was never acted on.

10. An index and verse praising a book written by his friend Tomás Tamayo de Vargas in 1635.

Don Philip comes across as a devoted son to his parents, a caring brother to his siblings, a devout Catholic, and a scholar rather than a man of war. In a letter to Fr Sinnot, attached to *The Compendium,* in the *Fragmenta,*[35] he says 'As I am at present, I hear neither the eloquence of Cicero nor the elegances of Valla or Manutius but the strange and barbarous language of sailors. The sound of cannon, which some dreadful person invented, for the destruction of the human race, is rattling in my ears. I am wielding the sword not the pen. How few are those that can excel in either; much less both.' He recognized that the divisions among the Irish were the cause of their defeat[36] and if they could unite that the English could be driven from the shores of Ireland. He hated the English for inflicting death, suffering and famine on his people and saw the English practising ethnic cleansing of the country.

## The Zoilomastix *(The scourge of Zoilos)*
The word 'Zoilomastix' is a word coined by Don Philip. Zoilos was a detractor of Homer and, among the ancient Greeks, a detractor of Homer (Homeromastix) was the lowest of the low, as Homer was held in God-like esteem by them.

---

35.  *Fragmenta* attached to the end of the *Compendium* F.264 v.
36.  *Hist. Cath. Hib. Comp.*, Tom. 3, Lib 1, cap. 5.

Mastix is the Greek word for a scourge or whip, so Don Philip indicates that he is the scourge of Zoilos and, in this case, Zoilos is Giraldus Cambrensis. In the first four books of *The Zoilomastix,* Don Philip attacks the '*The History and Topography of Ireland of Giraldus*'. He does not attack the '*Expugnatio Hiberniae*' of Giraldus which describes the conquest of Ireland by the Normans. In the fifth book of *The Zoilomastix,* the Zoilos is Richard Stanihurst. Don Philip attacks the appendix of the '*De Rebus in Hibernia Gestis*' of Richard Stanihurst, which deals with the Topography of Giraldus. I feel that when Don Philip started the Zoilomastix he only intended to attack the 'Topography' of Giraldus Cambrensis and that his attack on the '*De Rebus*' of Stanihurst in Book Five of the *Zoilomastix* was an afterthought. The first four books are devoted to attacking the Topography of Giraldus. At the end of Book Four, on page 307v, he says that he is submitting the book to the Inquisition for approval. In C.III of the later Retaliation, page four of the translation, entitled 'Stanihurst refuted these words,' he uses quotations from the Appendix of the '*De Rebus*' to refute Giraldus. This is because Stanihurst included in his Appendix annotations that were critical of Giraldus. Don Philip quotes these to bolster his arguments. In Book Five of the Zoilomastix Don Philip goes back to attacking Stanihurst as he had been attacking Giraldus in the first four books.

### The Manuscript

The manuscript of *The Zoilomastix* was lost for over 300 years. It was finished around 1626. We can deduce this from internal evidence in the manuscript. Don Philip, writing of Irish Ecclesiastics in the Fifth Charter of the twentieth retaliation (Book Three, Folio 244R), refers to Thomas Walsh, in Latin he calls him Thomas Valoysius and says that he returned to Rome from Spain 'Inde Romam adijit ubi casiliae archps. (i.e. *Archiepiscopus*) creatur' where he was made Archbishop of Cashel. This occurred on the 8th of July 1626. Don Philip does not mention the death of his great friend and collaborator Fr Richard Conway S.J. which occurred on the 1st of September 1626.[37] It would thus appear that the manuscript was finished somewhere between July and September 1626.

In January 1932, Professor Eoin MacNeill, Chairman of the Irish Manuscripts Commission, was contacted by the Keeper of Manuscripts at the University of Uppsala in Sweden, who informed him that there was in that library an unedited manuscript of a work by Don Philip O'Sullivan Beare. The title of the work was on the first page:

37.    MacErlean, John P., 'Richard Conway', *Irish Monthly*, 52, p. 47.

*Philippi O'Sullewani Bearri Hiberni vindiciae Hibernicae contra Giraldum Cambrensem et Alios vel Zoilomastigis liber primum, 2, 3, 4 et 5.*[38]

Page 2 attached to the manuscript says:

*De la libraria del Marques de Astorga en Madrid en el mesa de junio 1690. J.G. Sparwenfeldt.*

The manuscript passed into the library of the Marques de Astorga and from there was acquired by J.G. Sparwenfeldt in 1690 and then ended up in the library of the University of Upsala where it is catalogued as manuscript H.248. A third page attached to the manuscript shows:

*H.248*
*Biblioth Univ. Upsal.*
(coll. Sparwenfeldt)

The existence of the manuscript was known as a poem in Latin by, a man called Mendosa, a Portuguese, was prefixed to the *Patritiana Decas*,[39] a Life of St Patrick in ten books, written by Don Philip in 1629. Mendosa wrote:

*Mendacia Magna Gyracdi rejicit, et stolidus quae quae Stanihurstus habet notitia varia pulchrum, sermone politum Zoilomastix et dicitur illud opus.*

This translates as:

*'The great lies of Giraldus he refutes, and the lies contained in the works of stupid Stanihurst. That work adorned with diverse ideas and polished speech is called* The Zoilomastix.*'*

The manuscript of *The Zoilomastix*, written in Latin, differs from other writings of Don Philip. The *Compendium* and the *Tenebriomastix* start with a citation in flowery language dedicated to his Catholic majesty, the King of Spain. The opening of *The Zoilomastix* is incomplete, beginning 'add to [what has gone before] this the considered opinion which is attributed to Peter Lombard ...'. This would indicate that something has gone previously. There is no chapter heading to the first page and the first chapter heading on page two is Chapter Five. This indicates that Chapters One, Two and Three are missing and that the first pages of the manuscript probably represent the final page of Chapter Four. This makes the opening sentence of the manuscript, as we have it, rather unsatisfactory from the reader's point of view. In addition, the first retaliation we meet is *Retaliatio Posterior* which indicates that there was a

---

38.   See Appendix B.
39.   There is a good copy of this in Marsh's Library, Dublin. There is also a microfilm of it in University College Cork.

previous *Retaliatio Anterior,* the end of which can be seen in references to the Irish scholars Peter Lombard and David Rothe. I have decided not to discuss the technical aspects of the manuscript at great length in this publication.

The *Zoilomastix* is handwritten in a good hand, possibly by the author. The manuscript consists of 359 folios written on each side, giving in all 718 pages. Attached at the end of the *Zoilomastix* is a poem written by himself about thirty-seven of the Irish saints. He gives four lines to each saint. In addition to Patrick and Bridget, who come first, he includes many minor saints. There is a lot of crossing out in the manuscript with the new insertions acting as a correction for what is crossed out and in the same hand, as if the writer was seeking a better way of stating what was crossed out. It is unlikely that a scribe would make so many corrections. The corrections are put above what is crossed out or in the left or right margin with insert marks to show where he wants them inserted. This can make the reading of these passages difficult. The Tenebriomastix is written in a completely different hand, probably that of a scribe.

Fr O'Donnell transcribed the handwritten Latin into typed Latin for an MA at University College Dublin (UCD) in 1941 with a very good introduction. From this he published a book called *Selections from The Zoilomastix of Philip O'Sullivan Beare* which was brought out by the Dublin Stationery Office for the Irish Manuscripts Commission in 1960. In this, Fr O'Donnell takes excerpts in Latin from his MA thesis; however, he never produced a translation of the *Zoilomastix* into English.

*Why Don Philip wrote* The Zoilomastix

At the end of the sixteenth century a number of English writers wrote histories of Ireland which repeated the calumnies of Giraldus Cambrensis against Ireland. The revival of Cambrensis at the end of the sixteenth century and the early part of the seventeenth century was, if you like, a form of propaganda to justify the Elizabethan conquest and cruel suppression of Ireland and, in addition, to give the impression on the continent of Europe that the Irish were still as barbarous, without knowledge of their religion, treacherous and backward, as Giraldus said, so that the European powers, especially Spain but also France, would not send help to the Irish rebellions against the English. These writings in English included Buchanan, Camden, Campion, Chambers, Davies, Hanmer, Hooker and Stanihurst in Latin. They all wrote in similar vein about Ireland and reading their accounts becomes very repetitive. I consider Spenser as representative of this group, as he is the person with the greatest literary fame of these writers. He was also a planter and spent eighteen years there from 1580–1598 so was well acquainted with Ireland. All the writers I have detailed in Appendix A. A number of Irish writers took up their pens to refute derogatory writings about Ireland. These included Peter Lombard, David Rothe, Thomas Messingham, Stephen White, John Lynch, Geoffrey Keating, Hugh McCaugwell, Luke Wadding and John Wadding, and later Roderick O'Flaherty (see Appendix A).

Don Philip wrote the *Zoilomastix* attacking Giraldus and Stanihurst, as the work of Giraldus was still the accepted text on Ireland even though it was written in 1188. These works were republished in Frankfurt in 1602, by Camden. Of the Irish writers, Don Philip was the only layman among them, the rest all being clerics.

Don Philip's youth saw the destruction of the Old Irish system. With great cruelty his country was plundered, his people massacred on Dursey by Carew, and the O'Sullivans expelled. When his father, Dermot, and his first cousin, Donal Cam O'Sullivan Beare, joined Don Philip in Spain he would have been regaled about the sad end of Gaelic Ireland, after the Battle of Kinsale in December 1601. His hatred of the English exudes from the pages of the *Zoilomastix*. He considers English and heretic to be synonymous and lashes out at heretics and English with equal relish.

Don Philip in the *Zoilomastix* is acting as a propagandist, as both he and his relatives felt that the war against the English was not finished and that the Irish rebellion could still be revived with the help of Spain. The Spanish also hated the English and their Protestantism so Don Philip was feeding the type of propaganda that he felt would go down well with 'His Catholic Majesty,' the King of Spain. Don Philip is criticised for going over the top in his criticisms but in war no holds are barred.

### The Format of The Zoilomastix

Don Philip quotes a sentence from 'The Topography' of Giraldus Cambrensis and then proceeds to refute what Giraldus said. In Book I, Chapter I, in the Retaliatio Posterior (later retaliation), page three of the translation, Don Philip quotes the words of Giraldus:

> Gyraldus: *'Ireland indeed is uneven and mountainous, soft and watery, wooded and boggy, truly a country deserted without roads and wet.'*[40]

Don Philip indignantly goes on to refute these words. This format can lead to a lot of repetition throughout the book.

---

40.    Giraldus, *Topography*, I c.iv; See also *Zoilomastix* Retaliatio 10 ch. 8, p. 307.

In Book One he next proceeds to a description of Ireland and uses the quotations from twenty-one authors ancient and recent to confirm what a wonderful country Ireland is. He does not include any derogatory remarks. Giraldus, in describing Ireland, takes quotations from only Bede and Solinus. He gives a detailed description of the River Shannon, as Giraldus incorrectly states that the Shannon flows two ways, one into the Northern Sea and the other into the Brendanican [Western] Sea.[41] Likewise, with birds in Ireland, Giraldus refers to eleven types of birds[42] whereas Don Philip, in his description of Irish birds, refers to eighty different types of birds. Similarly, with fish, Giraldus records a few while Don Philip records many. I feel that Don Philip in thus recording far more than Giraldus is trying to prove that his refutation of *The Topography* is much more reliable than the writings of Giraldus.

The section of Book I of the *Zoilomastix* which deals with the natural history of Ireland, I am sure, will be of great interest to the readers. A lot of the descriptions of birds is taken from the 'Natural History' of Pliny the Elder and from Aristotle. Pliny had insatiable curiosity. He died in the eruption of Vesuvius on August 24th, 79 AD[43] when he went to get a closer view to satisfy his great curiosity and to rescue some of his friends.

The sections on insects and aquatic creatures also owe a lot to the 'Natural History'. The sections on eagles and owls comes from Pliny and are not clear, and I doubt if all the eagles mentioned lived in Ireland. A very interesting feature of this section is that the names of animals, birds, etc. are given in Latin, Greek, Spanish and Irish. I have translated these using the appropriate dictionaries. I am fortunate that my father was a native Irish speaker from Glengarriff and he taught me the names of animals, birds, fish, plants, etc. Coming from the country, he had a great interest in nature. It was a delight to read Don Philip's Irish names sounding the same as those I had learned as a youth from my father. In the Irish, the names are very often in the plural and are written phonetically. It is likely that Don Philip gets his descriptions in Irish from what he remembers when he left Ireland for Spain as a twelve-year-old in 1602, also that which he got from his father Dermot who probably would have continued to speak Irish to the large Irish retinue who surrounded Donal Cam, now Count of Birhaven, his Spanish title, and who was also a Knight of Santiago, one of the highest honours that a Spanish nobleman could aspire to. Don Philip's father, Dermot, was seventy years old when he left Ireland in 1602 and lived another thirty years to age 100. It is likely that he would have

---

41.  *Topography* Part 1 c.2. 5, *Topography*, 7–17.
42.  *Ibid*, Dist I, c.8, 9, 10.
43.  Pliny the Elder 23–79 wrote his *Natural History* of the Roman Empire which included Gall (modern France) and Britain, the flora and fauna etc. here would overlap that of Ireland to a large extent.

continued to speak Irish all his life so Don Philip would have been conversant with Irish from him. I asked Professor Terence O'Reilly, Professor of Spanish at University College Cork, to review the Spanish names and he felt Don Philip may have been writing them phonetically also.

The descriptions of the animals, birds etc. do not always coincide with modern ornithological definitions of the different species. For instance, in Ireland, birds may be called different names in different areas of the country and even in relatively local areas.[44] This makes the identity of some of the species difficult to define, purely on the Latin names but taking the four languages – Latin, Greek, Spanish and Irish[45] – and then Don Philip's description, it is possible to be sure in most cases exactly of which animal, bird, fish, etc. he is speaking. Fr T. O'Donnell in Appendix A of his Latin Selections from the Zoilomastix of Philip O'Sullivan Beare got Tomás de Baldraithe to give the Irish, Latin and English names in the natural history section of Book I of the Zoilomastix. He does not include the Greek and Spanish words. Another excellent work is *The Complete Guide to Ireland's Birds* by Eric Dempsey and Michael O'Clery (Dublin, 1993) as it gives the birds' names in Latin, Irish and English with pictures. This is a very good reference source.

When he has finished with the stones on page 131 of the translation, there is a sudden return to a description of birds, including birds which have been written up previously, e.g. the pheasant and sparrow. The writing in this section appears to be in a different hand as if somebody else inserted it later, but before the pagination was made, as the page numbers are in correct order. This hand appears on a small number of other occasions, usually in the margin, acting to correct something. This would indicate that some effort to edit the manuscript was made by a second person. In the original manuscript 'they discuss the purgatory of St Patrick' was at the top of Folio 53v[46] (page 132 of the translation), and the rest of the page was left blank. However, this has been filled in by the second hand writing on birds. For this reason, I finished off the section on birds and transferred 'they discuss the purgatory of St Patrick' to its present position on page 139 of the translation, as it fits in with the final section of the book.

---

44. Pearse, Patrick. 'Names of Birds and Plants in Aran', *Gaelic Journal*, ix, 1889, pp. 305–6.
   Forbes, Alex Robert, *Gaelic Names of Beasts (Mamalia, Birds, Fishes, etc.)*, 2 parts, Edinburgh, 1905.

45. I feel that Don Philip was showing off his erudition here to prove that he exceeded Giraldus in learning.

46. This is the pagination on the upper right hand corner of the page. A pagination in the lower right hand corner was probably the original but after Folio 29 it goes to folio 40 instead of 30 and the pagination on the upper * hand margin of the folio is to correct this.

The next three Books, II, III and IV continue to quote the chapter headings from the *Topographia* of Giraldus and then refute them. The fifth Book is a refutation of Stanihurst's *De Rebus in Hibernia Gestis*, which is a re-publication of the topographia but with explanatory notes that don't always agree with Giraldus.

# BOOK I

Iudicium libelli descriptionem Iberniae continentis, qui Petro Lombardo Ardmachae Archipontifici adscribitur, adde:
*post Bedam*, inquit, *longo tempore secutus est Sylvester Gyraldus Cambrensis, qui ab Honrico secundo Anglorum rego cum filio Joanne missus in Iberniam scripsit ex professo de hac insula, etc. Quamvis autem aliter quam historicum decet veracem, et modestum mordacior sit, et amarulentior in Ibernicam passim nationem quasi data opera, et alicubi in Scotiam captata occasione; tamen in describenda Iberniae regione commemorat multa, quae separatis (sicut debent) quos ille, et ubi inspergit aculeis, continent insignem, et veracem regionis huius commendationem.*

Donati Orruaki in *Refrigerio Antidotali* contra *Dempsterum* sunt haec verba. *Cambrensem adi Sylvestrem Giraldinum, tametsi in aliis multis nobis et toti genti iniquiorem.*

His, ni fallor, quam vanus, futilis, mordax, fallax sit histrio Gyraldus, liquet: cum duorum vel trium ore veritas colligi iubeatur.

# BOOK I

Add the considered opinion[1] which is attributed to Peter Lombard, Archbishop of Armagh, contained in a booklet describing Ireland.[2]

'*A long time after Bede*', he said, '*Sylvester Gyraldus Cambrensis was sent by Henry II, King of the English, with his son John to Ireland. Gyraldus wrote openly about this island, etc. Although a historian ought to be correct and restrained, Gyraldus however behaves otherwise. He is quite deliberately more biting and bitter everywhere when he writes about the Irish nation and anywhere when he gets the opportunity against Scotia.*[3] *However, Gyraldus also records many things in describing the region of Ireland which (setting aside, as is right, the barbs which he scatters upon it) contain outstanding and true commendations of this region.*'

The following are the words of Donatus O'Rourke in the *Refrigerium Antidotale* against Dempster, '*come to Sylvester Gyraldus Cambrensis although in many other things he is more unjust to us and to the whole race*'.[4]

From these, lest I am mistaken, it is clear how vain, futile, biting, deceitful, the actor Gyraldus is, since the truth should be collected from the evidence of two or three people.

---

1.   The extant text begins somewhere towards the end of the fourth chapter of the *Retaliatio Anterior,* hence the opening sentence alludes to immediately preceding text which no longer survives.

2.   Presumably Lombard's *De regno Hiberniae sanctorum insula commentarius,* Louvain, 1632; but neither the printed edition nor the extant manuscripts contain the quotation given by O'Sullivan, hence what is here marked as a quotation may simply be an extrapolation from the attitudes evinced passim by Lombard.

3.   That is, Ireland.

4.   *Hibernia Resurgens, sive Refrigerium Antidotale,* p. 60; a work by the Bishop of Ossory David Rothe, published under the pseudonym Donatus O'Rourk.

## C. V
### Quae Geraldi dicta sunt hic refellenda?

Nihilominus ut Iberniae splendor magis illustretur, et lectoris expectatione[5] satisfiat, illum fusius, et accuratius examinandum duxi. Eius autem singula deliramenta insectari taediosissimum esset. Illa duntaxat quibus Ibernorum gloriam obscurare, machinatus est refellere fert animus.

## Retaliatio Posterior
### Iberniae terrae laudes accumulantur

Insulam quidem ipsam quamvis miris laudibus saepissime praedicet, nihilominus eam etiam gravissime vituperavit, dum inter illius naturam describendam sic fuit locutus.

## C. I
### Gyraldi verba

*Distinctio I. Ch. 4.*    Gyraldus. *Ibernia quidem terra inaequalis et montuosa; mollis*
*Stanihurst, Appendix*   *et aquosa[6], sylvestris et paludosa, vere terra[7] deserta, invia sed*
*Ch. 5*      *aquosa.*

## C. II
### Horum verborum confutatio

*Epist. 92*      Philippus. Omnium calculo conceditur a Seneca sapiente probe traditum fuisse: *iniuria fortunae maiori saepe locum fieri.* Quod exemplo praesente clare probatur. Namque telus Iberniae Geraldi dente morsa ita multorum testimoniis nostraque veissima a descriptione illustris, et clara evadet, ut de illa vere diei possit 'calumnia laudi maiori novumquam locum fieri.'

---

5.    This is incorrectly written expectationi in the manuscript.
6.    'Mollis et aquosa' is inserted from the margin.
7.    An illegible word is crossed out before *deserta*.

## C. V
### What are the things that were said by Gyraldus that need to be refuted here?

Nevertheless, so that the splendour of Ireland may be better illustrated and satisfy the expectation of the reader, I have decided to examine Gyraldus more widely and more accurately. It would be very tedious to pursue every single one of his mad statements. I am moved to refute at least those things by means of which he has contrived to obscure the glory of Ireland.

## Later Retaliation
### The praises of the land of Ireland are heaped up

Although he frequently extols the island itself with very great and wonderful praises; nevertheless, he decried it in a most severe way when he spoke as follows, describing its nature:

## C. I
### The words of Gyraldus

*Distinctio I. Ch. 4. Stanihurst, Appendix Ch. 5*

Gyraldus. *'Ireland indeed is uneven and mountainous, soft and watery, wooded and boggy, truly a country deserted, without roads, and wet.'*

## C. II
### Refutation of these words

*Epist. 92*

*Philip.* It is commonly agreed that it was correctly stated by the philosopher Seneca *'that space is very often made for greater fortune by injury'* which is proven clearly by the present example.[8] For the land of Ireland, bitten by the tooth of Gyraldus, emerges, by the testimonials of many and by my very true description, so truly glorious and famous that it may be possible to say about

---

8.   This is from Seneca's *Epistulae morales ad Lucilium*, no. 91, paragraph 13: 'Saepe maiori fortunae locum fecit.'

Quem quidem laborem ego in praesentia non nisi istius
obtrectatoris convicio provocatus adire, institui.

# C. III
### Haec verba Stanihurstus reprehendit

Istum itaque suus acerrimus alioquin sequax Stanihurstus
non leviter reprehendit capite quinto Appendicis, quam ex
eius opere collegit, ita verba relata expendendo. *Alludit parum
commode ad istum vatis versiculum Psalmo 62. In terra deserta,
invia et inaquosa. Verum Iberniam non adeo desertam fuisse
eo tempore, etiam ipso Gyraldo teste, liquido apparet. C.1. ita
scribit. Poteram quidem, ut alii aurea forte munuscula, falcones,
et accipitres, quibus abundat insula, vestrae Sublimitati destinasse.
C.6. campos frugibus abunde vestiri docet. C.7. magnam vini vim
[-in Iberniam][9] asportari testatur. Passim in historia magnam
Ibernorum multitudinem in armis fuisse declarat. Quibus omnibus
in unum collectis, consequens est, Iberniam non fuisse desertam,
nisi illam terram desertam esse Gyraldus vult, quae aureis
munusculis abundat, in qua incolae agriculturae operam navant,
cum transmarinis mercatoribus[10] commercia habent, quae in
quavis insulae portione populis referta est.* His et aliis erroribus,
quos Stanihurstus fusius colligit, Gyraldum neque in veritate
constantem neque in mendacio strenuum esse, satis constat.

---

9.      O'Donnell in his transcription leaves out in Iberniam.
10.     Negocia habent is crossed out after mercatoribus.

it 'that space is sometimes made for greater praise by malicious accusation'.[11] For indeed I would not have undertaken this task at the present time if I had not been provoked by the reviling of that detractor Gyraldus.

# C. III
## Stanihurst refuted these words

Stanihurst himself, otherwise a very keen disciple, therefore did not lightly refute Gyraldus in the fifth chapter of the Appendix which he collected from his work by thus considering the words reported: '*He alludes, not aptly, to the verse of the prophet, in Psalm 62: "In a land deserted, without road and unwatered." Truly it appears clearly, even from the testimony of Gyraldus himself, that Ireland was not that deserted. In Chapter One he wrote as follows "I could, like others, have chosen for your sublime highness small gifts of gold, falcons, and hawks, with which the island abounds" – chapter 6. He tells us that the plains are covered abundantly with crops – chapter 7. He attests that a great amount of wine is imported into Ireland. He declares everywhere in his history that a great multitude of Irish men were under arms. When all of this is collected together, the result is that Ireland was not deserted, unless Gyraldus intends that land to be understood as deserted which abounds in gifts of gold and in which the inhabitants perform with diligence the work of agriculture, and carry on commerce with overseas merchants; a land which in all parts has a large population.*'[12] With these and with other errors, which Stanihurst has compiled more comprehensively, it is satisfactorily established that Gyraldus was neither constant in truth nor consistent in lying.

---

11.    This is an adaptation of Seneca's words, not a second quotation.
12.    Stanihurst, p. 225.

# C. IV
## Ea verba magis examinantur

Ut eius vero verba singulatim trutinem; si per terram inaequalem intelligit eam, quae tota omnino aequali, rasaque superficie non contineatur, aequalem in orbe terrarum quaerere, furor est: cum nulla sit regio perfectissime plana sin quae nullam habeat planiciem: significant, planicies laetas atque latas in Ibernia negare est etiam insania. Montes vero: sylvae: paludes ibi (fateor) sunt: sed et valles etiam, aperti campi, locaque sicca non desunt. Insulam: quae habitata, culta passim et pervia erat, desertam, et inviam appellare, summae fuit dementia. Aquis denique lacuum, fluminum, fontium, stagnorum sed quae incolis magis usui, quam incommodo sunt, eam abundare, non modo concedo, sed etiam contendo.

# C. V
## Gyraldus suis verbis quibus Iberniam mirabile laudat, refellitur

*Apud Stanihurst.*
*Appen c.6. C.7*
*Appendix apud*
*Stanihurst*

Verum ut cum hoc proteo vultus mutante tabulis agam consignatis suis verbis et ipsius amentiam confutabo, ac meam sententiam confirmabo.

Post repetita verba sic ait: *Planicies tamen habet per loca pulcherrimas. Rursus Gleba praepingui, uberique frugum proventu foelix terra est et foecunda. Frugibus arva, pecore montes, nemorosa feris abundant. Rursus Pascuis, et pratis, melle, et lacte, vinis etc. dives est insula. Rursus Fluminibus egregiis scinditur et rigatur, lacus quoque plurimos, piscosos, et grandes prae aliis terris, quas vidimus, haec profert. Marinis piscibus per omnia latera satis abundat. Flumina vero, lacusque, suis sibi innatis piscibus foecunda sunt. Accipitres, falcones et nisos*

## C. IV
### These words are examined in greater detail

But let me weigh his words one by one: if by reference to an uneven land, he thinks of a land that is bounded by a completely level and smooth surface, it is madness to search for this 'level' on earth, since no region is perfectly level; but if he means the sort of land that has no plain, it is also insane to deny there are pleasant and wide plains in Ireland. Truly, there are mountains, woods and marshes there, I admit: but there are also valleys, open plains and dry areas. It was the product of a deranged mind to call an island deserted and impassable, which was inhabited, cultivated far and wide, and has roads through it. I not only concede but even assert that it abounds in the waters of lakes, rivers, springs, marshes; but these are more a benefit to the inhabitants than an inconvenience.

## C. V
### Gyraldus is refuted by his very own words in which he praises Ireland in a wonderful way.

*Append. c.6 in Stanihurst. Appendix c.7 in Stanihurst*

However, to deal with this Proteus[13] who changes his shape, although he has put his seal on his writings, I will both refute his madness and confirm my own opinion. Next, after the words I have repeated, he says the following: '*However, it (Ireland) has beautiful plains in places.*'[14] AGAIN: '*The soil is very rich and the land fertile and fruitful in the abundant production of crops. The ploughlands abound in produce, the mountains with flocks of sheep, the woods with wild animals.*' AGAIN: '*The island is rich in pastures, meadows, honey and milk and vines, etc.*'[15] AGAIN: '*It is divided and irrigated by outstanding rivers; also it displays numerous fish-filled lakes which are large compared to other lands we have seen. It fairly abounds with sea fish on all sides. Truly, the rivers and lakes abound in their own native fish. This region produces hawks, falcons and sparrow hawks*

---

13.  Proteus: Shepherd of the underworld, herded the seals and sea creatures for Poseidon. He was also a prophet. He could change his shape at will, cf. *Odyssey* 4.
14.  Stanihurst, p.225; in fact this is from chapter 5 of the appendix.
15.  Ibid., p. 226; in chapter 6 of the appendix.

*prae aliis regionibus haec copiose producit. Aquilarum quoque non minorem copiam, quam alibi milvorum videas.*

*Apud Stanihurst*
*C.9 Append.*
Rursus *Terra terrarum haec omnium temperatissima. Non cancri calor exaestuans compellit ad umbras. Non ad focos Capricorni rigor urgenter invitat. Nives hic raro et tunc modico tempore durare, videbis. Ex omni tamen vento non minus Subsolari, Favonioque et Zephyro, quam Circio, et Boreali quandoque brumescit; ex omni quidem modice, ex nullo immoderate. Sicut aestivo, sic et hyemali tempore herbosa virescunt pascua. Unde nec ad pabula foena secari nec armentis unquam stabula parari solent. Aeris amoenitate, temperieque tempora cuncta tempescunt.*[16] *Aeris quoque clementia tanta est ut nec nebula inficiens, nec spiritus hic pestilens, nec aura corrumpens. Medicorum opera parum indiget insula. Morbidos enim homines, praeter moribundos, paucissimos invenies. Inter sanitatem continuam mortemque supremam nihil fere medium etc* Statim addit. *Item nemo unquam indigenarum hic natus, terram aeremque salubrem non egressus, ulla trium generum specie febricitavit. Sola vexantur acuta, eaque perraro.* Hae profecto Iberniae rarae insignes, et magnificae laudes sunt. Maioribus tamen adhuc eam Gyraldus praedicat, ita tradit: *Nulla tamen aeris virulentia, nulla // temporis intemperies vel sanos, et hilares hic contristat, vel etiam delicati capitis, cerebrum turbat.*

*Rursus: Quae videlicet Ibernia quanto a caetero et communio orbe terrarum semota, et quasi alter orbis dignoscitur; tanto rebus quibusdam solito naturae cursu incognitis.* *C.3 App. apud Stanihurst*

---

16.    The manuscript incorrectly writes tepescunt.

*more copiously than other regions. You may see also an abundance of eagles equal to that of kites elsewhere.'[17]*

*In Stanihurst c.9 Append.*

AGAIN: *'The land is the most temperate of all lands. The exhausting heat of the tropic of Cancer does not drive one to the shade. The cold of the tropic of Capricorn does not invite one urgently to the fireplaces. Here, you will see the snows rarely and then lasting a limited period of time. Winter now and then is marked by every wind, as much the east and west and south-west, as the north-west and north. Yet each, indeed blows in a moderate way, and none immoderately. Grassy pastures grow green in winter time, as in the summer. Thus they are not accustomed to cut hay for fodder and never prepare stables for the beasts. With the pleasantness and the mildness of the air, almost all seasons are moderately warm. The clemency of the climate is such that here neither fog taints, nor is the atmosphere unhealthy, nor is the breeze destroying. The island is in little need of the services of the doctors. You find very few ill people apart from those who are about to die. Between continuous health and final death, there is scarcely any mean'* etc. He immediately adds *'In the same way, no one of the natives born here who has not left the land and the healthy air, ever suffers from any of the three kinds of fever. They are troubled by the ague alone and that very rarely.'*[18] These truly are rare, outstanding, wonderful praises of Ireland. Yet Gyraldus marks it with greater praises still, and reports thus: *'However, no virulence of the air, no wildness of the weather depresses the healthy and cheerful, or even disturbs the brain of one suffering a nervous disorder.'*[19]

AGAIN, he says: *'What is clear is that to the extent that Ireland is far removed a great distance from the rest of the general body of the world and is regarded as though it were another part of the earth, to such an extent, by means of certain things, unknown in the usual course of* *C. 3 in Stanihurst*

---

17. Ibid., p. 227; in chapter 7 of the appendix.
18. Ibid., p. 233; in chapter 9 of the appendix – O'Sullivan has omitted a sentence, marked by 'etc'.
19. Ibid., p. 234; in chapter 9 of the appendix.

*Quasi peculiaris eiusdem naturae thesaurus, ubi insignia, et pretiosiora sui secreta reposuerit, esse videtur.*[20]

Rursus*: Inter omnia vermium genera solis non nocivis Ibernia gaudet. Venenosis enim omnibus caret: caret serpentibus, et colubris; caret bufonibus, et ranis; caret et scorpionibus; caret et draconibus. Rursus: Sed hoc stupore dignum occurrit, quod nihil venenosum aliunde advectum unquam continere vel potuit vel potest etc. Toxicum quoque similiter allatum mediis in fluctibus innata, malitia benignior aura privat etc. In tantum siquidem haec terra inimica veneno est, ut si aliarum regionum seu viridiaria, seu quaelibet alia loca pulvere ipsius aspergantur, venenosos vermes abinde procul exterminet. Corrigiae quoque terrae istius non adulterinae, sed verae et de coriis animalium, quae hic nata sunt factae contra serpentum, bufonumque morsus in aqua rasae, et potae efficax remedium ferre solent etc. Quinimo omnia quae de eadem insula sunt iuxta Bedae assertionem contra venena valent etc.*

*Quas igitur his comparabiles orientalis regio divitias iactat? etc. Omnes orientales pompas aeris nostri clementia compensamus etc. Terrae motus hic nunquam. Vix semel in anno conitruum audies. Non hic tonitrua terrent, non fulmina feriunt, non cataractae obruunt, non terrae motus absorbent, non leo capit, non pardus lacerat, non ursa devorat, non tigris absumit. Item non ciborum xenia*[21] *etiam ab hostibus facta ulla veneni suspicio reddit infesta. Non novercae privignus, non matronae quantumlibet offensae maritus toxicata pocula reformidat.*

---

20.   The manuscript writes videatur.
21.   The manuscript writes genia. Genius means taste or enjoyment of food, while xenium means a gift or a present.

*nature, it seems to be a special treasure trove where the aforementioned nature has deposited the most marked and the more valuable of her own secrets.'* [22]

*AGAIN: 'Among all the classes of creeping things, Ireland rejoices only in the harmless ones. It lacks all poisonous ones. There are no serpents or snakes. It does not have toads or frogs. It does not have scorpions or dragons.'* [23] *AGAIN: 'But this is a matter of wonder that it was and is unable to contain anything poisonous that has been brought from elsewhere. Also, in like manner, if a poison is brought in on the tide, the kindlier air deprives it of its innate evil, etc.* [24] *Indeed this land is so hostile to poison that if a garden or any other place in other countries is sprinkled with its dust it will banish poisonous worms far away from there.* [25] *Leather thongs also of this land which are not counterfeit but true and made from the skins of animals that are born here, if shaved into the water and drunk, usually offer an effective remedy against the bites of serpents and toads. Indeed, all things from the aforementioned island are effective against poison, according to the assertion of Bede, etc.* [26]

*What riches can the Orient boast comparable to these? etc.* [27] *We compensate for all the pomp of the Orient by the clemency of our air, etc.* [28] *There is never an earthquake here. Scarcely once a year will you hear thunder. Here thunders do not terrify, nor does lightning strike, nor do cataracts overwhelm, nor do the movements of the earth swallow one up, the lion does not capture you, the leopard does not tear you, the bear does not eat you, nor the tiger devour you. Nor does any suspicion of poison render gifts of food hostile, even when made by an enemy. The stepson does not fear the poisoned cup of his stepmother, nor the husband that of his wife, no matter how odious she is.'* [29]

---

22. Ibid., p. 222; in chapter 3 of the appendix.
23. Giraldus, *Topographia*, Dist. 1, chapter 28.
24. Ibid., chapter 29
25. Ibid., chapter 30.
26. Ibid. chapter 31.
27. Ibid., chapter 34.
28. Ibid. chapter 37.
29. Ibid. chapter 38.

*Rursus:*[30] *Hinc et venas metallicas, argenti mineralia, ferrique fodinas, pretiosa marmora, Parios et Lydios lapides, et nitentia alabastra, atque eximii pretii uniones orientalibus haud multum dispares, chrysoelectrum seu succinum, antimonii et aluminis venas, lapidis incumbustibilis pellucidas bracteas (quas alii speculares vocant) et alias multas tam reconditas quam exquisitas naturae opes solerti industria vel e visceribus elicias, vel in superficie colligas, ut nisi iners vel mentis inops, nemo despiciat tam divitem cornucopiam.* Haec Gyraldus qui cum tot verbis,[31] tantam fertilitatem, tantam amoenitatem, tantam temperationem, tantam salubritatem, tot, tantas, tamque mirificas, magnificas praestantes Iberniae laudes[32] exaggeret, ubi eam inaequalem, montosam, sylvestrem paludosam, aquosam, desertam, inviam asserit, vel benigna interpretatione declarandus, vel e lethaeo flumine bibisse lethargo laesarum mentium morbo fuisse correptus autumandus est: maxime cum alii neque pauci, neque silentio praetereundi scriptores ab Iberniae parte stent.

# C. VI
## Ibernia a variis authoribus describitur[33]

*Solinus*

*Julius Solinus* 'de Britannia verba faciens: multis (inquit) insulis, nec ignobilibus circumdatur, cuarum Iuverna ei proximat magnitudine etc. ita palubrosa, ut pecuaria, nisi interdum aestate a pastibus arceantur, in periculum agat // satietas.'

Pomponius Mela. 'Super Britanniam Iuverna est pene par *Mela* spatio, sed utrinque aequalis tractu littorum oblonga etc. adeo luxuriosa herbis non laetis modo sed etiam dulcibus, ut se exigua parte diei pecora impleant, et nisi pabulo prohibeantur, diutius pasta dissiliant.'

---

30.    From Rursus to cornucopiam is an insertion from the margin.
31.    After verbis Hae Gyraldus: qui cum is crossed out.
32.    After laudes, tot verbis is crossed out.
33.    Crossed out at beginning of C.VI. Ptolomaeus. 'Scotia eadem ac Ibernia Britanniae proxima spatio terrarum angustior, sed situ foecundior.' Ptolomaeus de Trib. Part. orb c.16 apud Magnesium.

*AGAIN: 'From here with skilled labour you can bring forth from the bowels of the earth, or gather on the surface, metal, silver, iron, precious marbles, stones of Paros and of Lydia and shining alabaster and large single pearls of enormous value not far different from the oriental ones, topaz or amber, veins of antimony and aluminium, clear thin sheets of the incombustible stone, (which other people call mica) and many other riches of nature, as hidden as they are searched for, so that nobody would despise so rich a horn of plenty, except someone dull or not in full control of his senses.'* Thus Gyraldus, with so many words, magnifies the great fertility, the great pleasantness, the great temperateness, the great healthiness of the country. He exalts Ireland with very many and very great astonishing attributes and outstanding praises, while he also declares that the island is uneven, mountainous, wooded and boggy, wet, deserted and without roads. Either he must be shown to have meant this kindly, or to have drunk from the river of forgetfulness, or to have been corrupted by a lethargic disease of damaged minds, especially when other writers (not a few and not to be passed over in silence) are on Ireland's side.

## C. VI
### Ireland is described by various authors

*Solinus*

*Julius Solinus,* describing Britain, says that: 'It is surrounded by many not ignoble islands of which Iuverna is the nearest in size to it, etc. It is so abounding in fodder that herds of sheep or cattle, unless they are kept away from the pastures during the summer, are in danger of over-eating.'[34]

*Mela*

*Pomponius Mela:* 'Beyond Britain, is Iuverna, almost equal in size but on both sides equal in extent in the length of shore, oblong, etc. It is so rich in foliage, not only lush but sweet, that in a small part of the day the cattle fill themselves so that unless they were kept away from forage they would burst if fed any longer.'[35]

---

34.   Solinus, *De mirabilibus mundi,* chapter 22 [Mommsen edition, 1864].
35.   Pomponius Mela, *De Situ Orbis,* book 3, chapter 6.

*Cornelius Tacitus.*[36] 'Siquidem Ibernia medio inter Britanniam    *Tacitus in vita*
atque Hispaniam sita, et Gallico quoque mari opportuna    *Agricolae*
valentissimam inperii partem magnis invicem usibus miscuerit.
Spatium eius, si Britanniae comparetur angustius, nostri maris
(i.e. Mediterranei) insulas superat. Solum, coelumque, et
ingenia cultusque hominum non multum a Britannia differunt.
Melius aditus, portusque per commercia et negotiationes[37]
cogniti.'

*Aethicus.* 'Ibernia insula inter Brittaniam, et Hispaniam longiore    *Aethicus*
ab Africa in Boream spatio Porrigitur etc. Haec propior
Brittaniae, spatio terrarum angustior coeli, solisque temperie
magis utilis a Scotorum gentibus colitur.'

*Paulus Orosius* presbyter Hispanus eadem fere verba habet.    *Orisuis L.I. Hist.*
'Ibernia insula inter Brittaniam, et Hispaniam sita longiore ab    *Contra Paganos*
Africo in Boream spatio porrigitur etc. Haec propior Britanniae,    *ch. 2 apud Petrum*
spatio terrarum angustior, sed coeli, solisque temperie magis    *Lombardum in*
utilis a Scotorum gentibus colitur.'    *descrip. Iberniae*

Haud dissimilia scribit divus Isidorus Hispalensis Episcopus.    *Isidorus L.14 c.6*
'Scotia eadem ac Ibernia proxima Britanniae insula, spatio    *Etymol (apud Petrum*
terrarum angustior, sed situ foecundior. Haec ab Africo    *Lombardum deleted)*
in Boream porrigitur, cuius partes priores Iberniam, et
Cantabricum oceanum intendunt; unde et Ibernia dicta: Scotia
autem, quod, ab (sic) Scotorum gentibus colitur appellata illic
nullus arguit.'

Venerabilis autem monachus Beda Anglo Saxo. 'Ibernia autem    *Bede in Hist. Lib I.*
et latitudine sui status, et salubritate, ac serenitate aerum    *Cap. 2*
multum Britanniae praestat, ita ut raro ibi nix plusquam triduana
remaneat: nemo propter hyemem aut fena sciet aestate, aut
stabula fabricet iumentis: nullum ibi reptile videri soleat. Nam
saepe illo de Britannia allati serpentes, mox ut proximante

---

36.    This whole passage is inserted from the margin.
37.    O'Donnell in his transcription wrote negotiatores which is incorrect.

*Tacitus in the life of Agricola*

*Cornelius Tacitus:* 'Inasmuch as Ireland, situated half-way between England and Spain, and convenient to the Gallic Sea, might join the strongest part of the empire with great mutual advantage. Its extent, if compared with Britain, is narrower, but it exceeds the islands of our seas [i.e. the Mediterranean]. The soil and weather, and the character and culture of the people, do not differ much from Britain; the approaches and harbours are better known through trade and business.'[38]

*Aethicus*

*Aethicus:* 'The island of Ireland extends between Britain and Spain a greater distance to the north from the south, etc. The island is nearer to Britain. It is a narrower land. It is more useful in the temperateness of its weather and temperature and is inhabited by clans of Scots.'

*Orosius Against Pagans, bk 1 ch.2. Quoted by Peter Lombard in Descpt Iberniae*

*Paulus Orosius,* a Spanish priest, says almost the same words: 'The island of Ireland extends between Britain and Spain a greater distance to the north from the south, etc. It is nearer to Britain and narrower in width; it is more advantageous in the temperateness of weather and heat. It is inhabited by clans of Scots.'[39]

*Isidore, Etymologies, bk 14, ch. 6.*

Not in dissimilar vein writes St Isidore, the bishop of Seville: 'Scotia or Ireland is the nearest island to Britain. It is narrower and its situation more fertile. It stretches to the north from the south and its headlands stretch out into the Irish and Cantabrian Sea. From that it is called Ibernia. On the other hand, it is called Scotia because it is inhabited by races of Scots. There nobody disputes this.'[40]

*Bede in his Lib 1. cap. 2 History, Bk 1, Ch. 2*

On the other hand, the venerable monk *Bede,* an Anglo-Saxon, says: 'Ibernia, indeed, because of latitude of its situation and in its healthiness and the serenity of the airs is more outstanding than Britain. Thus, it is rare there for the snow to last more than three days. Nobody would cut hay for the winter in summer time or would build stables for the cattle, no reptile is seen there.

---

38.   Tacitus, *Agricola*, chapter 24.
39.   Lombard, *Commentarius*, pp. 15–16.
40.   Isidore, *Etymologies*, 14: 6.

terris navigio, odore aeris illius adtacti fuerint, intereant. Quin potius omnia pene quae de eadem insula sunt, contra venenum valent. Denique vidimus quibusdam a serpente percussis[41] rasa folia codicum, qui de Ibernia fuerunt, et ipsam rasuram aquae immisam, ac potui datam talibus protinus totam vim, veneni grassantis, totum inflati corporis absumpsisse, ac sedasse tumorem. Dives lactis ac mellis insula, nec vinearum expers, pescium, volucriumque, sed et cervorum venatu insignis. Haec autem propria patria Scotorum est. Ab hac egressi (ut diximus) tertiam in Britannia Britonibus, et Pictis gentem addiderunt.'

*Apud Nicol. Serarium*  Author *vitae divi Killiana*. 'Scotia, quae et Ibernia dicitur, insula
*C.1 opusc. et*  est maris oceani foecunda quidem glebis; sed sanctissimis
*Messingham*  clarior viris; ex quibis Columbano gaudet Italia; Gallo ditatur Alemania; Killiano Teutonica nobilitatur Francia.'

*Theodoricus in vita*  *Theodoricus.* 'Insula sicut omni terrarum gleba fecundior, ita
*S. Rumoldi apud*  sanctorum gloriosa simplicitate beatior.'
*Sur. 1. Julii apud*
*Magnesium*

*In vita D. Cathaldi*  Bartholomaeus Moronus Italus. 'Cathaldus ex Ibernia oriundus:
*apud Magnesium*  quae insula in occiduo mari trans Britanniam sita est; insula quidem Britannia dimidia plus parte minor, sed par omnino et vel // ubertate agri, vel pecoris foecunditate atque etiam tepore soli, coeli clementia, et aeris serenitate nobilior.'

---

41.    O'Donnell writes serpentibus, in the MS as serpente.

For often when serpents are carried there from Britain, as soon as the ship nears shore when they come in contact with the air of Ireland, they die. Indeed, more, rather, almost all, things that come from that same island are effective against poison. Finally, we have seen that, when people were stricken by a snake, if the leaves of manuscripts from Ireland are scraped and the same scraping mixed with water and given in a draught to such persons, it immediately removes all the power of the spreading poison and takes away all the swelling of the body and settles it down. It is an island rich in milk and honey, nor is it lacking in vines, in fish and birds, but it is also famous for the hunting of stags. This land is the true fatherland of the Scots. When they went out from it (as we have said), they added a third race in Britain to the British and to the Picts.'[42]

*Nicholas Serarius, Opuscula vol 1. Also in Messingham.*

The author of *The Life of St Kilian:* 'Scotia, which is also called Ibernia, an island of the ocean which is indeed rich in ploughlands but more famous for its most holy men from among whom Italy rejoices in Columbanus, Germany is enriched by St Gall, German Francia is ennobled by St Kilian.'

*Theodoricus in a Life of St Romualdus quoted by Surius in July 1st quoted by Messingham.*

*Theodoricus.* 'The island, as it is more fertile in ploughland than all the world so it is more blessed in the glorious uprightness of its saints.'

*In Messingham, Life of St Cathaldus*

*Bartholomeus Moronus,* an Italian. 'Cathaldus was a native of Ireland. This island is situated beyond Britain in the western sea. The island is more than half the size of Britain but absolutely equal to it both in the fertility of the soil and the productiveness of the flocks and even more notable for the moderateness of the sun, the clemency of the climate and the serenity of the air.'

---

42.    Bede, *Historia Ecclesiastica*, 1: 2

Wallafridus Strab.[43] (in prologo *vitae S. Galli*). 'Ibernia insula,[44] ut scribit Orosius, inter Britanniam et Hispaniam sita longiore ab Africo in Boream spatio porrigitur. Haec propior Britanniae spatio terrarum angustior, sed coeli, solique temperie magis utilis: et ut supradictus Solinus testatur, ita pabulosa, ut pecua ibi, nisi interdum aestate a pastibus arceantur, in periculum agat satietas. Illic nullus anguis' etc.)

*(this inserted from L. margin)*. Munsterus Cosmographus de Ibernia. 'Mira coeli temperis et fertilitas terrae insignia.'

Georgius Bucananus Scotus. 'Coelum (scilicet Britanniae) Gallico temperatius; verum utroque mitius est Ibericum. Rursus Est enim ut ait Caesar coelum Britannicum Gallico temperatius. Ibernorum autem solum amoenitate, et coelum temperamenti aequabilitate Britannicum superat.'

Henricus Huntingdon.[45] 'Ibernia post Britannium omnium insularum optima est.' Rursus 'Mirabili igitur dono Deus hanc ditavit insulam multitudinemque sanctorum ad eius tuitionem in ea constituit; praeterea lacte[46] et melle ditavit, vinearumque non expertem, venatu piscium, et volucrum, cervorum et caprearum insignivit.'    *Lib. 1 Hist. Angl.    apud eund.*

Paulus Jovius episc. 'Ibernia nec dimidiam Britanniae partem aequare existimatur, cum ei par sit vel ubertate agri, vel pecoris foecunditate et quod Angli facile concedunt tepore soli, coelique,    *In descript Ibern.*

---

43.  (Ibernia proprior Britaniae spatio terrarum *ap Magnesium* Augustior sed coeli solisque temperie magis utilior) crossed out.

44.  Deleted in right margin: Val Strabo ap. Magnes-ium apud eundem Lib. 1 rerum scot. p2 apud eundum1.2 p.54 apud Mag.

45.  There is a blotch over the 'i' in Huntingdon in MS. O' Donnell decided it was an 'o' instead of an 'i'. There are two illegible words crossed out after Henricus.

46.  In MS it is lecte.

*Wallafridus Strabo.* In the prologue of *The Life of Saint Gall:* 'The island of Ibernia, as Orosius writes, is situated between Britain and Spain and stretches a longer distance from the south to the north. It is nearer to Britain and is narrower in the extent of territory but is more useful in the temperateness of its weather and of the soil. And, as the forementioned Solinus testifies in his previous statement, it is so abounding in fodder that the cattle there, unless they are kept away from the pastures during summer, are in danger of over-eating. There, there is no snake,' etc.

*Munsterus Cosmographus,* concerning Ireland: 'The temperateness of its climate is wonderful and the fertility of the land is outstanding.'

*Scotus p. 2 &*
*Messingham, book*
*2, p. 54*

George Buchannan a Scotsman: 'The weather (like that of Britain) is more temperate than that of Gaul. But the Irish weather is milder than both again. AGAIN: thus, as Caesar says, the British weather is milder than that of Gaul. The land of the Irish in its pleasantness and the weather in its uniform temperateness outshine those of Britain.'

*Book 1, History of the*
*English, quoted in the*
*above*

Henry of Huntingdon (thus): 'Ireland after England is the best of all the islands. AGAIN: Therefore by a wondrous gift, God enriched this island and he set up in it a multitude of saints for its protection. Besides he enriched it with milk and honey, made it not without vines, and made it famous for fishing and the hunting of birds, stags and roe deer.'

*Description of Ireland*

*Paulus Jovius,* bishop: 'Ireland is reckoned not to be half the size of Britain while it is equal to it in the richness of its soil or the fertility of its animals and a fact that the English easily concede it is more notable in the moderateness of the sun and the clemency of the weather and in the serenity of the air. A nation ignorant of luxury and untainted with foreign enticements, it is a nation that dislikes greatly the plough, avoids the mattock,

clementia, atque aeris serenitate nobilior. Gens ignara luxus, et peregrinis illecebris incorrupta: aratrum magnopere fastidit, logonem recusat, sudoremque omnem praeter bellicum refugit quo uno nihil fere praeter decus quaerendum existimat.'

Bartholo. Cassaneus. 'Ibernia insula Britanniae proxima spatio terrarum est illa angustior, sed situ foecundior, gentes habet ferocissimas, et bello aptissimas quae pulchro sunt corpore, et elato, membrisque robustissim is, ac colore candido, equosque parit, qui a natura suavissimo in cessu et quadam modulatione, veluti more regio deambulare videntur.' *In Catalog. Gloriae Mundi Part. 12 consid. 57*

Raymundus Marlianus,[47] in descriptione locorum Galliae, quae sunt apud Caesarem et Tacitum. 'Ibernia insula est ad Hispaniam, occidentemque solem sita, qua medium Angliae latus vergit etc. Pari spatio transmissis triginta millibus passuum in Iberniam appelles, ut in Britanniam. In hac sunt plures metropoles, et civitates episcopales, quam in Anglia. Eam quidem nominant Hirlandiam.'

*In Cosmographia*    *Petrus Apianus.* 'Irlandia est insula foecunda, et salubris, serpentibus, et venenosis reptilibus carens, equos mittit ferendis oneribus aptissimos.'

*In Historia suae Generis*    Gulielmus Neubricensis Anglus. 'Hanc autem singularem prae cunctis regionibus habet a natura praerogativam, et dotem ut nullum gignat venenatum animal, nullum reptile noxium. Quinimo certa, citaque mors eorum est ad primum Ibernici aeris attractum, si forte aliunde advehantur.[48] Porro quid quid avehitur contra venena valere, probatum est.'

*In Scoti Metaphys. tract de 1. Princ et Theorem*    Mauritius Ibernicus Archiepisc. Tuamen. 'Propter illud quod tangit exemplariter de ligno continuato lapidi in quodam fonte etc. Adverte, quod ille fons est in Ibernia, seu Scotia

---

47.    This paragraph is an insertion from the R. margin.
48.    O'Donnell writes 'advehitur' instead of 'advehantur'. In the MS he also omits 'porro quid quid avehitur' before 'contra venena valere'.

avoids all effort except warlike exertion in which alone it considers honour should be sought above all else.'

*In a catalogue Of the glory of the Earth Part 12 Section 57* *Bartholo Cassineus:* 'The island of Ireland is closest in distance to Britain. It is narrower than Britain in extent of territory but richer in its situation. It contains most ferocious people, most apt at war, who have beautiful and tall bodies and with the strongest of limbs and white in colour. The nation breeds horses who are, by nature, very smooth in their gait and seem to pace about in a certain rhythm as though in a regal manner.'

*Raymundus Marlianus,* in a description of the places of Gaul in Caesar and Tacitus: 'Ireland is an island situated towards Spain and the setting sun which verges on the middle of the side of England (etc.). Having travelled the same distance of 30 miles as you would to Britain (from the continent) you will arrive in Ireland. In this island there are more towns and episcopal cities than in England. The English call it Hirlandia.'

*Cosmographia* *Petrus Apianus:* 'Ireland is a fertile island and healthy. It is lacking in serpents and poisonous reptiles. It exports horses most suitable for carrying loads.'

*In the history of his own race* William of Newburgh, an Englishman: 'It has, however, above all regions this special sign and gift from nature, above all regions, that it has a gift that it begets no poisonous animal, and no noxious reptile. In truth, a certain and swift death is their fate at the first touch of Irish air, if by chance they are brought from elsewhere. Furthermore, whatever is taken out of Ireland has been proved to be an antidote to poisons.'

*In the Metaphysics of Scotus concerning first principle and theorem* Maurice O'Fihely, Archbishop of Tuam: 'Concerning that which he touches upon in an exemplary way concerning the wood united to the stone in a certain well, etc. Take note that

Maiore seu Herlandia in partibus septentrionalibus insulae.
Signum enim affixum illi aquae ad limum usque inclusive,
perato determinato tempore, reperitur tripartitum, continuum
tamen. Nam pars supereminens aquae remanet lignum, pars
media ferrum, et pars infima lapis; et hoc non mirum est
respectu // aliorum plurium, quae in hac insula inveniuntur.
Nam ibi locus est, in quo nullum animal mori potest, quamdiu
ibi deget. Sunt et arbores ex se generantes aves fore anserum
sylvestrium quantitate, sunt alia loca, in quibis cadavera
illico resolvuntur; et alia in quibis perpetuo manent illasa.[49]
Est etiam tota terra immunia ab omnibus venenosis, imo
tamquam remedium contra ea, quocumque feratur. Plura alia
quaerantur apud antiquos. Qua re conciliatorem 10. Particula
Problematum in Commento Problematis. 7. Taceo incolarum
virtutes, et mores, ne assentari nostratibus videar. Animi
tamen nobilitate, et armorum vehementia caeteros excellere,
asserere non ambigo.'

Fr. Bartholom. Anglicanus. 'Ibernia oceani est insula in Europa *De propriet. rerum*
Britanniae insulae vicina: spatio terrarum angustior, sed situ *L. 15*
est foecundior. Haec ab Africo in Boream porrigitur, ut dicit
Isodorus lib. 14. Eius partes priores in Iberum et Cantrabricum
Oceanum tendunt. Inde enim ab Ibero oceano est Ibernia
nominata. Est autem insula frumentariis copiis uberrima;
fontibus et fluviis irrigua; pratis (etc.). Sylvis amoena; in metallis
foecunda, et gemmifera. Nam ibi gignitur lapis sexagonius,
scilicet iris, qui soli appositus facit in aere coelestem arcum. Ibi
etiam invenitur lapis qui gagates dicitur. Ibi margarita candida
reperitur. Quantum ad salubritatem coeli est Ibernia regio valde
temperata. Modicus est enim ibi excessus, vel nullus in frigore
vel calore. Lacus habet mirabiles et fontes. Nam ibi est lacus,
in quo si per aliquod spatium longum palus ligneus infigatur,
pars quae est terra infixa convertitur in ferrum, pars, vero, quae

---

49.    This marvel is widely commented on by seventeenth century
       authors.

this well is in Ireland or Scotia Major or Herlandia in the northern parts of the island. For the wood fixed in the water right up to the mud inclusively, after a fixed time has passed, is found in three parts, but continuous. For the part that is standing out of the water remains wood, the centre part iron, and the lowest part stone and this is not to be wondered at, considering the many other things found in this island. For there is a place there in which no animal is able to die, however long he will live there. There are trees that bring forth birds almost equal in size to wild geese, there are other places in which corpses melt away on the spot, and others in which they remain forever undamaged. For the whole country is immune from all poisons, rather it is a remedy against them wherever it is carried. Many others may be sought among the ancients. For which, see the Conciliator, on the 10th clause of the *Problems* of Aristotle, in the commentary on problem 7. I stay silent on the virtues and the character of the inhabitants lest I appear to flatter my own people. However, in their nobility of mind and the power of their arms, I positively assert that they excel others.'

*Concerning the property of matter, bk 15.*

Fr Bartholomew Anglicanus: 'Ireland is an island of the ocean in Europe, a neighbouring island of the island of Britain, narrower in extent of territory, but by situation more fertile. This stretches to the north from the south, as Isidore says, book 14. The nearest parts tend mostly towards the Iberian and the Cantabrian Ocean. It is named Ibernia from the Iberian Ocean. Indeed, it is an island most fruitful in abundance of corn: it is irrigated with springs and rivers; it is enhanced by meadows and woods, rich in metals and gem-bearing. For there is found a six-sided stone, namely the *iris*[50] which, when placed opposite the sun, makes a rainbow in the air. There also is found a stone which is called jet. There is found the white pearl. As regards healthiness of climate, Ireland is a region truly temperate. Excess of heat or cold there is moderate or non-existent. It has marvellous lakes and fountains. For there is a lake there in which if, over a long

---

50.    P1.37.9.62 pure six-sided prismatic crystal.

est in aqua in lapidem, sed pars, quae supra aquam est manet lignum.'

*Corylus arbor ferens*
*avilanas*

Est et alius lacus, in quem si projicias virgas coruli, convertuntur in virgas fraxini, et e converso si virgas fraxineas immerseris in coruleas mutabuntur. Sunt ibi loca, in quibus cadavera mortuorum nunquam putrescunt, sed ibi semper manent incorrupta. Est etiam in Ibernia quaedam insula parva, in qua homines non moriuntur, sed quondo nimis senio afficiuntur, ut moriantur, extra insulam deferentur. In Ibernia nullus est serpens, nulla rana, nulla aranea venenosa. Immo tota terra adeo contraria est venenosis, ut terra inde delata, et sparsa serpentes perimat, et bufones. Lana etiam Ibernica, et animalium coria fugant venenosa; et si serpentes, vel bufones navigio in Iberniam deferantur, statim moriuntur. Multa alia sunt mirabilia in illa.'

*In descrip. Iberniae*

Camdenus Anglus de Ibernica loquens. 'Pecorum hic infinita multitudo; ovium etiam greges plurimi, equi item optimi, quibus non idem, qui caeteris in cursu gradus; sed mollis al terno crurum explicatu glomeratio. Apum tanta multitudo, ut non solum in alvearibus, sed arborum etiam truncis, et terrae cavernis reperiantur etc. Utque uno // verbo dicam, sive terrae foecunditate, sive maris, et portuum opportunitatem, sive incolas respicias, qui bellicosi sunt, ingeniosi, corporum lineamentis conspicui mirifica carnis mollitie et propter musculorum teneritudinem agilitate incredibili, a multis dotibus ita felix est insula, ut non sale dixerit Gyraldus,[51] naturam hoc Zephyri regnum benigniori oculo respexisse.'

---

51.     Having examined a number of recensions of Giraldus' Topograpia Hiberniae I could not find this quotation.

period of time, a stake of timber is driven into the bottom of the lake in the part which is fixed in the ground, is converted to iron but the part which is in water is converted to stone and the part which remains above the water remains wood.

*Hazel tree bearing avilanas[52]*

And there is also another lake in which if you throw twigs of hazel they are converted into twigs of ash and conversely if you immerse twigs of ash they are converted to hazel. There are places there in which bodies never rot but remain always uncorrupted. Also, there is in Ireland a certain small island in which men do not die. But when they become too old they are carried out to die outside the island. In Ireland there are no serpents, no frogs, no poisonous spiders. Indeed, the whole land is so hostile to poisonous animals that earth when brought from there and spread about kills serpents and toads. Irish wool and the skins of animals ward off poisonous animals and if serpents or toads are brought on a ship into Ireland, straight away they die. There are many other wonders in Ireland.'

*In his Description of Ireland*

*Camden:* an Englishman, speaking of Ireland: 'Here there is an infinite number of cattle, also very many flocks of sheep and the best of have not the same step as others in their gait, but a gentle rotting movement stretching the near and far side legs alternately.[53] There is a great multitude of bees; they are found not only in beehives but even in the trunks of trees and in caverns of the earth etc. And, in a word, whether in the fertility of the soil, or of the sea, and whether you look at the convenience of the ports or you consider the inhabitants, who are warlike, intelligent, outstanding in the configuration of their bodies in the wonderful softness of their flesh, and of an incredible agility because of the suppleness of their muscles, this island is so fortunate with many blessings that Gyraldus rightly said that nature looked on this kingdom of the west with a very benign eye.

---

52.  Hazelnuts; presumably the Spanish word was more common in this period, reflecting the most common source of the food.
53.  This is a quotation from Pliny, 8: 166 – in fact it refers to Spanish, not Irish, horses.

*Joannes Davis Anglus* regis Jacobi procurator in Ibernia. 'Durante   *Apud Magn.*
meo servitio in Ibernia, omnes huius regni provincias diversis
itineribus iuridicae circuitionis obivi. Ubi interim observavi
clementiam aeris, et bonam temperiem ubertatem soli, iucundos,
et commodos habitandi situs, tutos, ac largos portus ac fidas
navigantibus stationes, et traiectum in omnes occidui orbis partes
liberrimum, longos fluviorum navigabilium tractus, et spatioses
lacus, ac stagna mediterranea; nullis per universam Europam
secunda, opulentam piscationem, et aucupium multiplex
omnigenarum volucrum, corpora, mentesque, indigenarum raris
et extraordinariis naturae dotibus praedita etc. Descriptio terrae
Canaan quae habetur Deutor. 8. optime quadrat in omnem
Iberniae partem. Terra rivorum, aquarumque, et fontium, in
cuius campis, et montibus erumpunt fluviorum abyssi; terra
frumenti et hordei; terra lactis, et mellis ubi absque ulla penuria
comedes panem tuum, et rerum abundantia perfrueris.'

Equidem vereor, ne tot authores, qui tellurem Iberniae
summis laudibus praedicant, referendo, lectori[54] satietatem,
ac taedium pariam. Quis igitur vel Iberniae regionis peritus,
vel in cosmographorum, et historicorum lectione versatus, vel
commerciorum septentrionis gnarus, nisi mentis incompos vel
maledicendi libidine actus Iberniam tanta coeli temperatione;
aeris clementia; agri ubertate; montium, sylvarum, planiciei
utilitate, portuum commoditate; terrestrium, volatilium,
aquatilium copia; metallorum opulentia; mirandarumque
rerum varietate insignem, claram inclytam turpibus mendaciis
audebat obscurare?

---

54.     O'Donnell writes lectorem but lectori in the MS is correct.

John Davies, an Englishman, the Attorney-General of King James in Ireland: 'During my service in Ireland I traversed all the Provinces of this kingdom in diverse journeyings of the law circuit. There sometimes I observed the clemency of the air and the good temperature and the fruitfulness of the soil, the pleasant and suitable sites for living, the safe and numerous ports, and the safe anchorages for shipping, and the very free passage to all parts of the western world. I observed long stretches of navigable rivers, and spacious lakes and inland ponds second to none throughout all Europe. I observed the rich fisheries, many bird catching methods of all types of birds, and that the bodies and minds of the inhabitants have been endowed with rare and extraordinary gifts of nature, etc.. The description of the land of Canaan which is contained in Deuteronomy 8 best fits every part of Ireland. It is a land of rivers, and of water, and springs, in whose plains and mountains burst forth bottomless rivers. It is a land of corn and barley, a land of milk and honey where you will eat your bread without any want, and you will enjoy to the full an abundance of things.'

Indeed, I fear that my referring to so many authors who speak about the land of Ireland with the greatest of praises may make the reader satiated and bored. Who is there who has personal experience of the land of Ireland or is learned in the writing of geographers and historians or experienced in trade with the north, unless he is out of his mind or motivated by a desire for abuse, who would dare to denigrate Ireland, a land which is outstanding, famous and renowned for a rarity of wonderful things – the temperateness of the climate, the mildness of the air, the fertility of the land, the usefulness of the mountains, woods, plains, the suitability of the ports, the abundance of animals, birds, and fish, and the richness of its metals?

## C. VII
### Iberniam author incipit describere

Caeterum a nullo horum, ac ne ab omnibus quidem fortasse, quos memini, omnes eius res memoria dignae comprehenduntur.

*Tomo I Lib. I Comp.*     Quamobrem, ut domestici scriptoris verbis exactius delineantibus, ob oculos lectori spectanda proponatur, quamquam eam alibi fusius descripsi, hic tamen breviore quidem[55] authorum citatione perstrictam,[56] sed copiosiore memorabilium rerum mentione exaggeratam pingere putavi.

## C. VIII
### Iberniae universalis descriptio

Itaque Ibernia multis nominibus Iberia, Iera, autoris descriptio (10) terna, Iuverna, Scotia maior et antiquior,[57] // Irlandia insula sacra,[58] insula sanctorum, Britannia minor, Ogygia nuncupata, gleba fertilis, coelo temperata.[59]

*Iberniae nomina (10) vide comp. tomo 1 l. 1 Fertilitas Temperatio situs*

Oceano undique cincta, inter septentrionalem plagam, occidentemque solem sita est.[60] Ubi austrum versus maxime tendit, sub quinquagesimo primo plus minus; ubi ab arcticum coeli cacumen proxime adcedit, sub quinquagesimo septimo altitudinis gradu iacet,[61] quo loco longissimus dies, sole cancrum attingente, horis circiter octodecim constare perhibetur.

---

55.     After quidem 'verborum narratione' has been crossed out.
56.     Perstrictam is inserted from L. margin.
57.     Maior et antiquior inserted from L. margin.
58.     From sacra to ogygia is inserted above the line.
59.     After temperata in the text some illegible words are crossed out.
60.     Crossed out in the margin is insula sanctorum and the rest is illegible.
61.     Iacet is inserted above the line and crossed out a the end of the line.

## C. VII
### The author begins to describe Ireland

But the whole amount of Ireland's features worthy of recall is not included by any of those authors whom I have mentioned, and perhaps not even by all of them together.

*Compendium bk. 1, h. 1.* For this reason, so that, with the words of a native writer making a more accurate sketch, Ireland may be put forward for examination before the eyes of the reader, although I have described it elsewhere at greater length, here, however, I have thought to depict it by curtailing quotations of authors but by providing a more copious account of notable things.

## C. VIII
### A general description of Ireland

*The author's description. The names of Ireland 10. See Compendium vol. , book 1. Fertility, Mildness, Situation* Ireland has then been called by many names, Iberia, Iera, Ierna, Iuverna, Greater Scotia, and Older Scotia, Irlandia, the sacred island, the island of saints, lesser Britain, Ogygia. It has fertile soil, and a temperate climate.

It is surrounded on all sides by the ocean, it is situated between the northern zone and the setting sun. Where it stretches most to the south it is around the fifty-first latitude more or less, where it extends nearest to the northern pole, it lies under the fifty-seventh line of latitude. In this place the longest day, when the sun touches the tropic of Cancer, is established to last for around 18 hours.

Ab oriente Britanniae eam partem, quae dicitur Anglia, unius diei secunda navigatione adsitam, partem quae Albion,[62] a septentrione fere et Scotia minor, atque nova vocatur, propinquiorem respicit.[63]

Ab Euronoto Galliam habet vix plus duorum dierum marino itinere remotam. Hispanias tridui navali cursu dissitas alibi noto, sive Africo in Aquilonem[64] ventum occurrentes spectat, in Cantabricum Galletiacumque pelagus intenta. Pari spatio a septentrione ab Islandia insula, ultra Albionem sita, distat. Ab occidente ab America orbis infimi parte nova inventa diffusissimo mari dirimente, longissime abest.

Figura ovata, vel lenticulari ab Africo in Aquilonem[65] maxime    *Figura* porrecta Italis sexcentis passuum millibus longa, lata trecentis iudicatur.

## C. IX
### Iberniae particularis descriptio

In quinque regiones praecipuas, dimidiamque regionem    *Regiones* dividitur. Hae sunt, duae Momoniae ad meridiem vergentes, Ultonia ad septentrionem tendens, Lagenia ad Orientem, Connachta ad Occidentem solem strata; et inter has Mithia dimidia regio non omnino media collocata.

*Episcopatus Oppida*    In Momoniis est archiepiscopale solium Casilia, cui episcopi novem ex tredecim facti, unus qui (1) Lismoriae, Manapiaeque, praeest, alius (2) Clueniae[66] Corcachiaeque praefectus, alius (3) Ardfertae, et Achadeoae praepositus, item (4) Rosensis et Kennechensis unus, tum similiter (5) Lomnachensis Cathaensisque[67] unus, (6) Laonensis, (7) Imilachensis, (8) Finibricensis, (9) Finuriensis obtemperant.

---

62.    Albion et scotia minor inserted from L. margin.
63.    A septentrione fere inserted from the R. margin.
64.    Sive Africo in aquilonem inserted from margin.
65.    Ab Africa in aquilonem inserted from left margin.
66.    Clueniae inserted from L. margin.
67.    Cathaensisque inserted from R. margin.

On the east it faces back at that part of Britain, which is called Anglia, situated a single day's sail away with a favourable wind; roughly to the north, it faces the part which is called Albion and lesser Scotia and New Scotia, which is nearer. To the south south-east lies Gaul, scarcely two days of ocean sailing away.

Extending into the Cantabrian and Galician sea, Ireland faces Spain, three days sail away in the south or south-west, where it faces north. It is separated an equal distance on the north from the island of Iceland, which is situated beyond Scotland. On the west, it is very far distant from America, the part of the earth's end most recently discovered, separated off by a very wide sea.

*Outline*

Its outline is oval, like the shape of a lentil. Its longest extent is from south to north. It is judged to be six hundred Italian miles long and three hundred miles wide.

# C. IX
## A description of Ireland by regions[68]

*Regions*

It is divided into five principal regions and one half region. These are two Munsters, situated towards the south, Ulster stretching to the north, Leinster to the east, Connaught lying towards the west and among these Meath, a half a region situated not exactly in the centre.

*Episcopal towns*

In Munster is the archepiscopal See of Cashel [Casilia], which nine bishops answer to, appointed out of thirteen dioceses. One is in charge of (1) Lismore [Lismoria] and of Waterford [Manapia], another is in charge of (2) Cloyne [Cluenia] and Cork [Corcachia] and another in command of (3) Ardfert [Ardferta] and Aghadoe [Achadeoa]. Likewise there is one bishop for (4) Ross [Rosensis] and Kinneigh [Kennechensis], then, similarly, one for (5) Limerick [Lomnachensis] and

---

68. Ken Nicholls of the History Department, UCC, an expert in place names has kindly given his assistance with this chapter.

*Doria, quae dorium
et Lucus vocatur*
In Ultonia est Ardmachae archipontifex. Ei pontifices novem ex decem creati, unus qui Dunum Conerethamque regit; Mithiensis, *et* Killmorensis, Cluanensis, Ardachadensis, Clocharensis, Rapetensis, Doriensis, Dromorensis obediunt.

In Lagenia Dubhlinniae archipraesul imperium tenet, cui pontificatus quatuor Killdariensis, Fearnensis, Lachlensis et Ossyrgiensis subduntur. In Connachta Tuamia archiflamen ditionem habet, cuius authoritatem sequuntur pontifices sex Duavensis, Cluanfertensis, Magionensis, Alfinensis, Archadensis, Aladensis. Abbatias, monasteria, et alias ecclesiasticorum sedes scribendo percurrere, taediosissimum fuisset. Oppida nonnulla moenibus cincta; et non modo indigenarum, sed etiam externorum frequentia; commerciisque non obscuros populos recensebo, plura tamen praetermittens. In Lominiis sunt // Manapia, Corcacha, Lomnacha, Dungarbanum, Omagaunipons, Moala, Killmuchelloga, Pratum Mellifluum, Kensalia, Ochella, Danguina, Tralia: In Connachta Galuea, Athlonia, Balathrriega: In Ultonia Culrathinna, Pontana, Lur, Stradbalia, Rupes Fergusia, Carlinnum:

Scattery Island [Cathaensis], (6) Killaloe [Laonensis], (7) Emly [Imilachensis], (8) Kilfenora [Finibricensis], and (9) Fore [Finuriensis].

*Derry which is also called Dorium and Lucus*

In Ulster is the Archbishop of Armagh. Nine bishops appointed from ten dioceses answer to him: (1) Down and Connor [Dunum Coneriamque], (2) Meath [Mithiensis], (3) Kilmore [Killmorensis], (4) Clonmacnoise [Cluanensis], (5) Ardagh [Ardachadensis], (6) Clogher [Clocharensis], (7) Raphoe [Rapotensis], (8) Derry [Doriensis],[69] and (9) Dromore [Dromorensis].

In Leinster, the Archbishop of Dublin holds the power, under whom are four bishops of (1) Kildare [Killdariensis], (2) Ferns [Fearnensis], (3) Leighlin [Lachlensis], and (4) Ossory [Ossyrgiensis]. In Connaught, the Archbishop of Tuam [Tuamia] holds sway, whose authority six bishops follow, (1) Kilmacduagh [Duavensis], (2) Clonfert [Cluanfertensis], (3) Mayo [Magionensis], (4) Elphin [Alfiniensis], (5) Achonry [Achadensis], and (6) Killala [Aladensis]. It would have been tedious to list the Abbeys, Monasteries and other ecclesiastical houses. I shall review some towns surrounded by walls, and communities which are not unknown for large populations, both native and foreign, and for their commerce. However, I shall pass over more. In Munster are (1) Waterford [Manapia], (2) Cork [Corcacha], (3) Limerick [Lomnacha], (4) Dungarvan [Dungarbanum], (5) Bandon Bridge [Omagunipons], (6) Mallow [Moala], (7) Kilmallock [Killmuchelloga], (8) Clonmel [Pratum Mellifluum], (9) Kinsale [Kensalia], (10) Youghal [Ochella], (11), Dingle [Danguina], (12) Tralee [Tralia]. In Connaught: (1) Galway [Galvea], (2) Athlone [Athlonia], (3) Athenry [Balathrriega]. In Ulster: (1) Coleraine [Culrathinna], (2) Drogheda [Pontana], (3) Lurgan [Lur], (4) Dundalk [Stradbalia], (5) Carrickfergus [Rupes Fergusia], (6) Carlingford [Carlinnum].

---

69.    Doire, Oak Grove. Dorium and Lucus, a grove.

In Lagenia Dubhlinna;[70] Asia, Killchenia, Rosa, Lacus Gormanus, Magnus Portus, Killmintana. In Mithia Molendinum, Taxus, Athbuia. Alia oppida, pagos et innumera propemodum castella, non est quid numeremus. Alias res memorabiles ab hominum oblivione vindicemus. Hic haud desunt ardui montes, herbidi tamen, fontibus liquidis rigati, et vertice nonnusquam paludem continentes: ob id armento, et pecore abundantes.

Hic sylvae densissimae, caeduae, fructiferae, feris frequentatae, occurrunt. Hic fretum circumiacens piscibus scatens, Hispanos, Gallos, Anglos, Belgos piscatores accit. Hic plurimi portus navibus a tempestate defendendis confugio tutissimo. Hic piscosi tam lacus navigabiles, amoenas insulas medias cingentes, quam pellucidi modo per planicies, modo per nemora tranquillo cursu delati fluvii ad capiendum salmones, tructas, anguillas, et id genus pisces invitant. Hic piscatus tam marinus, quam fluviatilis et hamatilis et saxatilis,[71] tum retis, sive nassae iactus, tum tridentis citus delectat. Venatio, et aucupium non parum voluptatis adfert. Pingues colles sese leniter tollentes, apertaeque planicies fertili gleba multa genera frumenti ferunt. Horticulti, saltusque fructus, atque salubres plantas gignunt.

*Montes*     Singulorum rerum nonnullae nominandae: ac primum aliquot montes Momoniarum Brandanus, Cruachus, Deadus, Galthius, Mangarta Crotus: Connachtae Nevinnus, Riegus, Divi Patritii mons, Golbuinnus: Ultonia Argillus, Borichus, Divi Donardi mons; ubi et eius extat sacellum, atque fons.

---

70.     Inserted from R. margin.
71.     Et hamatilis et saxatilis inserted from R. margin.

In Leinster: Dublin [Dubhlinna] (1) Naas [Asia], (2) Kilkenny [Killchenia], (3) Ross [Rosa], (4) Wexford [Lacus Gormanus], (5) Arklow [Magnus Portus], (6) Wicklow [Killmintana]. In Meath: (1) Mullingar [Molendinum], (2) Trim [Taxus], (3) Athboy [Athbuia]. There are other towns, villages and nearly innumerable castles that there is no reason to count. Let us rescue other notable matters from the forgetfulness of man. Here, there is no lack of lofty mountains, however they are grassy and irrigated by clear springs and in some places containing bogs at their summit; because of this, they abound in cattle and sheep.

Here are found very dense woods suitable for felling, fruit-bearing, frequented by wild beasts. Here, the surrounding coastal water lies teeming with fish, and brings in Spanish, French, English and Belgian fishermen. Here, many ports act as a very safe refuge for ships that need to be protected from a storm. Here navigable lakes surrounding pleasant islands in the middle, and very clear rivers carried on their tranquil course, now through plains, now through groves full of fish invite the catching of salmon, trout, eels, and that type of fish. Here, angling gives delight, sea fishing as well as river fishing, and fishing with hooks, rock fishing, the swift throwing of the net or the wicker basket, or of the tri-pronged spear. Hunting and fowling bring no small delight. Rich hills rising gradually and open plains produce many types of corn from their fertile soil. Gardens and woods bring forth fruits and health-giving plants.

*Mountains*

Some individual features should be named. First some **mountains** of Munster: Brandon [Brandanus], Hungry Hill [Cruachus], Deadus [Deadus], Galtee [Galthius], Mangerton [Mangarta], Slieve gCrot [Crotus]; of Connaught: Nephin [Nevinnus], Gamh [Riegus], Croagh Patrick [Divi Patritii Mons], Ben Bulbin [Golbuinnus]; of Ulster: Errigal [Argillus], Mourne Mountains [Borichus], Slieve Donard/Mountain of St. Donard [Divi Donardi Mons], where both his shrine and well are seen.

*Sylvae*

Sylvae Momoniarum Dosia, Vallis aspera, Diamhraca, Dromphininus. Connachtae Calria, Nea, Artacha: Ultoniae Conkenia Ultacha, Barlinna, Moera: Lageniae Dunana, Muluria, Osilia.

*Portus*

Portus Momoniarum Lomnecha, Fiennida, Danguina, Dumbea, Cruacanum, Konsalia, Corcacha, Ochella, Dungarbanum, Manapia: Connachtae Sligecha, Muega, Irrisa, Galvea: Ultoniae Pontana, Stradbalia, Carlinnum, Rupes Fergusia, Feurus, Sulinus, Luachrus, Ernius: Lageniae Dubhlinna, Killmantana, Portus Magnus, Lacus Gormanus, Belehaca.

*Lacus*

Lacus Momoniarum Lenus, Longus, Siadus, qui in Croti montis cacumine locatus Praecissis rupibus cingitur. Eo a Divo Lachtino apes multas e Connachta pulsas fuisse, et ex favis, quos in rupibus condunt, aestuante sole liquefactae cerae rivos in subjectum lacum manara fertur. // Lacus Connachtae Riechus, Corbius, Keius, Gilia: Ultoniae Ernius, Beathacus, Neachus, Cuanus et insignis Divi Patritti purgatorio Deargus.

*Flumina*

Flumina Momoniarum Brocus, Duilius, Felius, Labuinnus, Magnus, Mandus Mangus, Maigus, Mulcherdus, Lius, Siurius: praeterea Cronsechus, Jonsechus, Febhnosachus: Connachtae Sligecha, Bullia, Muigus, Galuea:

The **woods** of Munster: Deise [Dosia], Glengarriff [Vallis aspera], Diamhraca [Diamhraca], Dromfineen[Dromphininus]; of Connaught: Calry [Calria], Nea [Nea], Airteach [Artacha]; of Ulster: Glenconkeen[Conkenia], Kilultagh [Ultacha], Kilwarlin [Barlina], Moira [Moera]; of Leinster: Doonaun [Dunana], Glen Malure [Muluria], Osilia [Osilia].

**Ports** of Munster: Limerick [Lomnecha], Fenit [Fiennida], Dingle [Danguina], Dunboy [Dumbea], Crookhaven [Cruacanum], Kinsale [Kensalia], Cork [Corcacha], Youghal [Ochella], Dungarvan [Dungarbanum], Waterford [Manapia]; of Connaught: Sligo [Sligecha], Moy [Muega], Erris [Irrisa], Galway [Galvea]; of Ulster: Drogheda [Pontana], Dundalk [Stradbalia], Carlingford [Carlinnum], Carrickfergus [Rupes Fergusia], L. Foyle [Feurus], L. Swilly [Sulinus], Ardbara Luachrus], L. Erne [Ernius]; of Leinster: Dublin [Dubhlinna], Wicklow [Killmantana], Arklow [Portus Magnus], Wexford [Lacus Gormanus], Ballyhack [Balehaca].

**Lakes** of Munster: Loch Lene [Lenus], Long Lake [Longus] and Lough Muskry [Siadus] which is situated in the summit of Slieve gCrot and is surrounded by sheer cliffs. There it is held that many bees were driven by St Lachtinus from Connaught and that from honeycombs which they hide in the cliffs, when the sun is hot, rivers of molten wax drip into the lake below; Lakes of Connaught: L. Ree [Riechus], L. Corrib [Corbius], L. Key [Keius], L. Gill [Gilia]; of Ulster: Erne [Ernius], Gartan [Beathacus], Neagh [Neachus], Strangford [Cuanus] and Loch Derg [Deargus], famous for St Patrick's Purgatory.

The **rivers** of Munster: Brocus [Brocus], Deel [Duilius], Feale [Felius], Laune Labuinnus], Blackwater [Magnus], Bandon [Mandus], Maine [Mangus], Maigue [Maigus], Mulkear [Mulcherdus], Lee [Lius], Suir [Siurius], also Cronechus [Cronsechus], Funchion [Jonsechus], Febhnosachus [Febhnosachus]; of Connaught: Sligo [Sligeacha], Boyle

Ultoniae Ernius, Deargus, Banna, Moernus, Finnius, Ennachus, Jescachus, Leninnus: Lageniae Bearba, Feorius: Hi et ex Momoniis post longa spatia dimensus est, labens Siurius, omnes tres iuxta Manapiam coeuntes: ibique Momoniis a Lagenia disterminatis, communi ostio mare influunt: Mithiae Bonnius et Niger.

# C. X
## Sinonni Amnis descriptio

*Sinonus*[72]

Omnium vero totius Iberniae fluminum non modo maximus, sed cum aliarum regionum nobilissimis conferendus est Sinonnus.

*Sliebh an ierann*[73]

Fons eius in Ferreo Monte oritur, loco viridi, iuncosoque septus, ore quidem angustus, ita tamen altus, ut eius profunditas nulla unquam bolida potuerit explorari. Hinc Sinonnus originem ducens in meridiem spectat. Principio fonte rivus modicus emittitur, qui convenis[74] aquis stipatus per montis dorsum decem millia passuum porrigitur, donec sub illius radice

*Lochellinn*

excurrat in Ellinnum lacum longitudine circiter quatuor, latitudine duorum millium passuum patentem: aliquot insulas, unam Coenobio divi Francisci celebrem, alias parvis fanis sacratas cingentem, ex eo lacu magno diffusus incremento minoribus fluviis receptis ita convalescit, ut navium patiens, longinqua loca Connachtae irriget. Inde lenissimo cursu labens longis excursibus a Connachta Lageniam dividit. Rursus extenti itineris intervalla emensus Connachtam, et eam Momoniarum partem, quae dicitur Urmonia, interfluus scindit. Hinc inter Tomoniam et Dutharam Dergertum lacum octo millibus passuum longum efficit.

---

72.  He incorrectly writes Soninnus crosses it out and corrects it to Sinonnus.
73.  Sliebhanhierinn is a phonetical spelling, the H before ierinn should be omitted.
74.  Convenis is written in the MS but should be convenientibus.

[Bullia], Moy [Muigus], Galway [Galvea]; of Ulster: Erne [Ernius], Derg [Deargus, Bann [Banna], Mourne [Moernus], Finn [Finnia], Eany [Ennachus], Easu [Jesachus], Lennan [Leninnus]; of Leinster: Barrow [Bearba], Nore [Feorius] – when these rivers, and the Suir, flowing out of Munster having travelled a great distance, come together, all three of them, near Waterford, there they divide Munster from Leinster, and flow together into the sea by one mouth. The rivers of Meath are the Boyne [Bonnius] and the Blackwater [Niger].

# C. X
## Description of Shannon River

*Shannon*

Truly, of all the rivers of the whole of Ireland the Shannon river is not only the greatest but can be compared with the most noble of other regions.

*The Iron Mountain*[75]
*(Sliabh An Iarainn)*

Its source arises in the Iron Mountain, hedged in by a green and rushy place. It is narrow at the mouth but so deep, that its depth has never been able to be sounded by the plumbline. The Shannon, originating from there, it flows towards the south. At first, the river is relatively small as it emerges from its source until augmented by converging streams, and running ten miles along the ridge of the mountain, it runs out at the foot of the mountain

*Lough Allen*

into Lough Allen, a lake that is around four miles long and extends two miles in width, which laps a number of islands, one famous for a monastery of St Francis, others sanctified by small shrines. It pours out from that lake, having been increased greatly by the addition of smaller rivers, and grows so strong that it is navigable for ships and irrigates the far-off regions of Connaught. From there, gliding in a most smooth flow, in long stretches, it separates Connaught from Leinster. Again, having run a long distance, it divides Connaught from that part of Munster which is called Urmhugha as it flows between them. From here, between Thomond and Arra, it forms Lough Derg eight miles long.

---

75.    The Shannon actually rises in Culcach Mountains at Shannon pot.

*As Danainn*

Ex quo effusus[76] usque ad Asdaninnum placidissime manat. Hic per teli iactum ita saxis occursantibus asperatur, ut naves quidem sed non pisces transitu intercludat. Hoc vadoso canali relicto iterum inoffensus, et suo more navigabilis means postquam millia passuum circiter sex processit, internum spatium ad insulae faciem amplectitur. In ea insula oppidum validis moenibus cingentibus est situm: cum quo aliud muris etiam firmatum a meridie sublicio ponte coniungitur: utrumque in unius speciem redactum Lomnecha dicitur, urbs situ pulcherrima et amoenissima. Ab altera oppidi parte cum amnis mari infusus iterum coit, portus patentissimus et navibus defendendis accommodissimus succedit. Amnis a fonte, unde primum auspicatum esse docuimus, ad hoc usque oppidum per millia passum plus ducenta, si omnes anfractus sequamur, dulcem haustum incorrupto sapore detinet, plerumque etiam navigabilis. Alias insulas ambit, alios[77] lacus facit, olorum natatione frequentatur, salmonum, tructarum, anguilarum variorumque piscium copia scatet: vel aperta planicie, vel loetis nemoribus atque sylvis cingitur: in his aquilae, columbae cum multiplici aliarum avium genere nidos aedificant: cervorum, leporumque greges minime desunt. Igitur tum piscatus, tum venatio, tum aucupium magnam otiosis voluptatem affert: et satietatem quam unius exercitationis frequentia parit, trium

*Fontes*

delectabilium commutatio vicissitudoque aufert. De fontibus quos hic liquido, gelido, tersissimo liquore gratissimos passim invenias, dicere longissimum fuisset.

---

76.    After effusus 'placidissime manat' is crossed out and put at the end of
       the sentence.
77.    Alios is inserted above the line.

As it pours out of this to Ard na Crusha, it flows very smoothly. Here, over the distance of a spear throw, it is made so rough by the presence of rocks, that it interferes with the passage of ships but not of fish. When this shallow area of the channel is left behind it is again uninterrupted and after its usual wandering but navigable course has advanced for about six miles, it embraces an area within it in the form of an island. In that island a town is situated surrounded by strong walls. To this another town also strengthened by walls is joined on the south by a bridge on piles. The two taken as a single unit are called Limerick, a city most beautiful and pleasant in its situation. On the other side of the town, where the river joins together again and pours into the sea, there is a port, very wide and most suitable for protecting ships. The river, from its source from where we have learned it began, right up to this town, is more than 200 miles long if we were to follow every bend. It remains fresh to drink and has a sweet taste, and is also mostly navigable. Sometimes it runs around islands, sometimes it makes lakes. It is crowded with swans swimming, and a multitude of salmon, trout, eels, and various kinds of fish abound there. It is surrounded by either open plains or pleasant groves and woods. In these, eagles, doves, with many types of other birds build their nests while herds of deer and hares are plentiful. Therefore, the fishing, the hunting, and the fowling bring great pleasure to the leisured classes: and great distaste that the frequent use of one of these pursuits brings the alternation and change of the three types of amusements takes away. It would have taken too long to tell of the springs which you may find everywhere, most pleasing in their clear, cold, clean water.

# C. XI
## Iberniae animantia terrestria

*Animalia*

Ad animantium considerationem properemus. Ut initium ab eo, quod caeteris regem, atque dominum Deus optimus maximus praefecit, sumamus:

*1. L. Homo*
*G. ἄνθρωπος*[78]
*H. Hombre*
*I. Iber. Duini*

Haec insula homines statura plerumque proceros, et elegantes, facie non deformes, caloris, frigoris, sitis, famis patientes, adversis rebus invictos, atque magnanimos bello per-quam aptos, litterarum studio flagrantes, christianae pietatis tenacissimos, haeresis, novorumque dogmatum hostes acerrimos procreat. Quod a nobis alibi, quia fusius est demonstratum, et demonstrabitur, hic diutius immorandum minime duco.

*Quadrupedia*

Ad bruta calamum transferamus, primum quadrupedum, quae cum lumine tellus communis nutrix alit, natura speculaturi, exordio a notissimis quibusque (quoad fiere potest) petito.

*2. L. Canis*[79]
*G. Κύων*
*H. Perro*
*Ib. Madri*

(Igitur canis animal gratissimum, quia et in custodia, et in venatione latratu, vel cantu signum dat, a canendo nomen accepit. Ex eius generibus),[80] quae haec insula gignit, qui tantum vel aedes latratu custodiunt, vel dominos comitantur, infimi dicuntur. His tamen utiliores meo quidem iudicio non sunt, quos ob parvitatem corporum[81] festivitatemque foeminae nobiles in doliciis habent, sive villosi, sive nuper invecti implumes.

---

78.　　G. Anthropos is deleted in L. margin and ἄνθρωπος is inserted from R. margin. This pattern is repeated throughout this whole section of the natural history. The word in Roman letters is deleted and replaced by the Greek letters.

79.　　In L. margin before Canis, Madri and Canes are crossed out, before Κύων cyan is crossed out. Ib. Madri is plural for dogs and is phonetically spelt. It should be Madraí. This is common throughout this section of Book I and will not be referred to again. The singular for dog in Irish is madra.

80.　　Deleted at beginning of this paragraph 'ex animalis gratissimi canis' and from igitur to generibus is inserted from R. margin.

81.　　After corporum a word is crossed out? Lepide.

# C. XI
## The land animals of Ireland

*Animals*

Let us proceed to the consideration of animals; let us make a beginning from that one which God, the best and greatest, set up as king and lord over the rest.

*Man[82]*

This island produces men tall in stature and elegant with well-formed countenance, capable of putting up with heat, cold, thirst and hunger, unconquered in adverse circumstances, high-minded and well fitted for war. They are zealous in their application to learning, most tenacious of christian piety, bitter enemies of heresies and revolutionary dogmas. Since I have demonstrated this more fully elsewhere and will demonstrate it again, I do not consider that we need to dwell on it further here.

*Four-legged animals*

Let us transfer our pen to the brute animals, first inquiring into the nature of the quadrupeds – which (along with the sun) the earth, the mother of all, nourishes – and beginning with the most important ones, in so far as is possible.

*Dog*

Therefore, the dog, a most pleasing animal because in guarding and in hunting it gives the signal by its barking or singing (cantu), it gets its name from the way it sings (*canendo*.)[83] From the types of dogs which this island produces, those which guard only the house with their barking or go as companions to their masters are said to be the lowest types. In my opinion though, less useful than these dogs are the ones the noble women have as their favourites, because of the smallness of their bodies and their cheerfulness, whether shaggy, or bald ones, recently imported.

---

82. At this point in the manuscript, O'Sullivan Beare adopts the procedure of listing the Latin, Greek, Spanish, and Irish names for the things which his text is describing.

83. Cano, here in Latin means 'to sing'. This form of etymologising goes back to the Classical author Varro, whose *De lingua latina* established the principle of connecting the origin of the word to the nature of the thing.

Qui vel ferarum vestigia rostro scrutentes, inquisitorem ad latibula perducunt, vel aves[84] in flumine, lacu: proximoque maris sinu a venatore iactu stratas in terram deducunt, generosi iure censentur. Qui vero feras et cursu superant et viribus, ferocitateque interimunt, eos longe nobilissimos esse fatemur: maxime cum capite magno, rostro rotundo, ore grandi, dentibusque acutis, oculis nigris, auribus parvis et tenuibus pectore amplo, ventre gracili, cauda // falcata commendentur. Hos Iberni Cuas[85] vocant. Elegantia corporum, pernicitate et animositate omnium canum orbis longe praestantissimos.[86] Horum quidem minores lepores, vulpes, meles vincunt: maiores cervos etiam, lupos, apros interficiunt: omniumque Iberniae quadrupedum sunt principes longe velocissimi, animosissimi, ferocissimi. Ita quem horum unus comitetur, eum etiam desertum, et solitarium nulla fera per Iberniam totam audebit aggredi. Omnes in dominos fidissimi sunt. Pro eis generosi nonnulli saepe dimicarunt, illisque mortuis nonnumquam cibe abstinuerunt, donec inedia fuerunt consumpti. Omnium alutae in chirothecas leniendis, dealbandisque manibus aptas, et alios usus minime spernuntur.

Cum canibus iungendae sunt domesticae Iberniae feles, a *3. L. Felis* summa cautione cati, quasi cauti, quoque numinatae: quae *G. αἴλουρος* fulgentium oculorum acie ron minus noctu quam interdiu *H. Gato* pollentes, mirum est, quanto silentio, quanto speculatu, quam *I. Cait* levibus vestigiis in mures, et etiam aves exsiliant.

Si canes demas, mansuetorum quadrupedum fidissimi sunt equi. *4. L Equus* In hac insula generosi, animo etiam, velocitateque excellunt, *G. ἵππος* gravi, fastuoso[87] regioque (ut ita dicam) mollisimo tamen *H. Cavallo* atque lenissimo gressu incedunt, crura mirifica glomeratione *Ib Each capaill*[88] explicantes. Eorum maximus quisque proeliando assignatur, quod robore, animo, ferocitate praecellit.

---

84.   After aves a word is crossed out. Feras?
85.   Cua in right margin is not crossed out.
86.   O'Donnell writes 'praestantissos'.
87.   O'Donnell writes 'fatuoso'.
88.   Eacha is crossed out before 'equus'. G. Hippos is crossed out and replaced by 'Ittitos' in L. margin.

Those rather which search out, by scent, the tracks of wild beasts and bring the hunter to the hiding place or which bring to land birds shot down by the hunter, in the river, lake and adjacent sea bay, are rightly considered noble. Those which truly excel wild beasts in running and strength, and which kill them with ferocity, I say are by far the most noble. These are most to be recommended when they have a big head, a round nose, a large mouth, sharp teeth, black eyes, small and slender ears, a good size chest, a slender belly and a sickle shaped tail. The Irish call them hounds (*cuas*). For elegance of body, swiftness and courage, of all the dogs of the world they are the most outstanding. The smaller of these catch hares, foxes, and badgers, but the large ones even kill deer, wolves, and wild boar. Of all the four-legged animals of Ireland they are the leaders by far in speed, courage and ferocity. Thus the man whom one of these accompanies, even when he is deserted and alone, no wild beasts through the whole of Ireland will dare to attack. They all are most faithful to their masters. On their behalf many often nobly fight and when their masters die sometimes they abstain from food until they die of hunger. The tanned hide of all of them is prized when made into gloves, for softening and whitening the hands and for other uses.

Along with dogs we must associate the Irish domestic cats. *Cat* From their excellent caution they are also called *cati*,[89] as it were *cauti:* because of the sharpness of their flashing eyes which are powerful no less at night than during the day, they are a marvel how quietly, how watcfully, and with what lightness of foot they jump on mice and even birds.

If you exclude dogs, horses are the most faithful of the *Horse* domesticated quadrupeds. In this island they are noble and they excel also in spirit and in swiftness. They progress with a heavy, proud, regal (as I might say), but also with a gentle and very smooth step, stretching out their legs in a wonderful trot. The biggest of them is assigned to fighting as it stands out in strength, courage and ferocity.

---

89.    A play on the word *catusa'*, male cat and *cautus'*, sharp.

*Virg. 3 Georg.*

'*Si qua sonum procul arma dedere*
*Stare loco nescit micat auribus, et tremit artus*
*Collectumque premens volvit sub naribus igne*'

Minoribus nobiles iterfacientes vehuntur. In genere tertio sunt caballi incessu, et animositate degeneres: ob id oneribus portandi, aliisque servilibus operibus mancipati.

*5 L. Boves*[90]
*G. βοῦς*
*H. Buey*
*I. Bo*

Boum videas hic armenta densissima. Eorum mares tauri aspectu generositatem praeseferunt fronte torvi, auribus setosi, dimicationem cornibus poscentes, arenam in altum prioribus pedibus spargentes, de imperio vaccarumque coitu dimicant.

*Virg. 3. Georg.*

'*Illi alternantes multa vi proelia miscent*
*vulneribus crebris: lavit ater*[91] *corpora sanguis*
*Versaque in obnixos urgenter cornua vasto Cum gemitu.*'
Certamine inferior fugam arripit:

*Virg. ibid.*

'*Victus abit, longeque ignotis exulat oris Multa gemens: ignominiam,*
*plagasque superbi Victoris, tum quos amisit inultus amores*'
Vaccae fecundissimae lactis, butyri, casei uberem proventum suppeditant castrati boves alteri cibo alteri agriculturae adscribuntur.

*6. L. Ovis*[92]
*G. Ὀις*
*H. Oveja*
*I. Cuiri*

Hic oves aliae albae, aliae nigrae, aliae gilvae non modo lac, caseum, sed etiam quotannis bis tonsiles[93] bina vellera in hominum vestes

*7. L. Capra*[94]
*H. Cabre*
*G. Ἀιξ*
*I. Gabhair*

Caprae quoque discolores lacte salubri, saepeque in medicinam. Physicorum decretis epoto, et unde caseus, butyrumque coagulatur, manant.[95]

---

90.   Bus is crossed out before Βοῦς.
91.   O'Donnell writers alter.
92.   G. Ois crossed out and ὄις inserted.
93.   Crossed out after 'tonsiles' illegible word followed by 'aliorum ovium naturam'. Ater 'bina' valner is cossed out and replaced by 'vellera'.
94.   Gabhair is crossed out.
95.   Crossed out in R. margin 'sunt qui putent a danis primum in Iberniam post Bedae tempores una cum pupis et vulpibus'.

*Virg. 3 Georg. 84–85*     *'If some arms give off a sound far off he does not know how to stand in one place, he pricks his ears and his limbs tremble. And snorting, rolls beneath his nostrils the gathered fire.'*

Nobles, on a journey, ride the smaller ones. In the third type there are horses which are ignoble in pace and courage: because of this, they are sold for the purpose of carrying loads and other servile works.

*Bull*     Here you will see huge herds of cattle. Bulls are the male ones. They display their nobility by their looks, fierce of forehead, with bristling ears, inviting battle with their horns, scattering sand up high with their front legs. They fight for the control and the covering of the cows.

*Virg. 3. Geor. 220–223*     *'The bulls in turn join in battle with great force, with frequent wounds. Black blood washes the bodies and with a mighty bellow their horns, turned, are driven against the attacking foe.'*
The inferior betakes himself in flight from the contest.

*Virg. ibid. 225–227*     *'The conquered goes away and is exiled on unknown shores far away, bemoaning greatly the ignominy and the blows of the haughty victor; and the loves he has lost unavenged.'*
Cows are very fruitful in milk and butter, and they produce an abundant supply of cheese. Bullocks are assigned, some to food, some to agricultural work.

*Sheep*     Here, some sheep are white, others black, some light yellow. Not only do they supply milk and cheese but also every year, because there are two shearings, two fleeces, for mens' clothes.

*Goat*     Speckled goats also produce health-giving milk which is often drunk as a medicine on physicians' prescription, and from this, milk is thickened into cheese and butter.

*8. L Sus*
*G. ὗς vel σύς*
*H. puerco*
*I. muic*

Sues pinguissimae, maxime iuxta silvas, ubi glandibus pascuntur, sunt.

Hinc caro bovina (quae totius // orbis terrarum optima habetur) oxina, caprina, suilla: hinc victuli, agni, hoedi, et aegris utiles refrendes abundant: hinc equina, bovina, ovina, caprina, vitulina, agnina, hoedina coria partim in calceos, partim in chirothecas superfluunt.

*9. Cervus[96]*
*G. ἔλαφος[97]*
*H. Ciervo*
*I. Carrfhie*

Ne vero feras silentio praeterire videamur, hic pinguium cervorum confertissimos greges inventias. Est animal hoc simplex omnium rerum miraculo stupens, ita ut equo, sive bucula propius accedente, hominem iuxta venantem vel non cernat, vel mirans, non fugiat. At urgente cane homines obvios aggredientes cornibus male mulctare solent: Hi namque sunt ramosis cornibus defensi.

Minores his damas, quae cornibus simplicibus, amplioribus *L. Damae* tamen, et in fronte flexis muniuntur, hic conspicias nonnunquam *I. fiegh fearba* fortiter pugnare.

*Lib 5.*

Unde Martialis cecenit.
*Frontibus adversis molles concurrere damas Vidimus, et fati sorte iacere pari.*
Utrorumque et carnes comeduntur, et tergora in thoraces, et alios usus vertuntur.

---

(note 95 *cont.*) There are people who think that they were first brought into Ireland, after the lifetime of bede by the Danes along with wolves and foxes!!).

96.   Carrfhiagh is crossed out before L. Cervis in R. margin and G. Elaphos is crossed out in R. margin and ελαφος is inserted from the L. margin

97.   Inserted in L. margin after ἔλαφος is the following in a different hand 'deinde dam taini est animal habet cornua ampla'. I could not make out the second and third words but it seems to refer to the fallow deer, an animal with large horns.

Pigs are found to be fattest near woods where they feed on acorns.

From these animals come the flesh of heifers (which is held the best of all the the world), oxen, goat, and pigs. From these, calves, lambs, kids, and young animals useful for sick people abound.[98] From these, the hides of horses, cows, sheep, goats, calves, lambs, and kids, are in great supply: some are made into shoes, some into leggings, others into gloves.

Lest we appear to pass over wild animals in silence you will find here most dense herds of fat deer. It is a guileless animal dumbstruck by the strangeness of everything, so that if a horse or a heifer approaches close by, either it does not see the hunter nearby, or being amazed, it does not flee. When they are pressed by dogs, with their horns they tend to injure grievously men who come against them and attack them. For these deer are protected by branching horns.

Smaller than these are the fallow deer protected by simple, but bigger horns, bent over their foreheads; here, sometimes, you may see them fighting bravely.

Thus, Martial sang:
*We have sometimes seen the soft fallow deer running with foreheads poised for attack and lying by the equal lot of fate* [i.e. dead].
The flesh of both types is eaten and their skins are turned into breastplates and other useful things.

---

98.    'Nefrendes' are young animals that have not yet developed teeth; they were assumed to be good for the ill.

Apros, sive silvestres, ferosque sues (bellicosa, et ferocia bruta, cumque provocantur, inermibus timenda) hic magna pinguedine ferunt[100] fuisse.

10. L. Aper[99]
G. ἑρραός
H. javali
I. muic fhiain

Meles, alio nomine, taxus, vulpe maior hic alitur: in cavernis dormit: natura pugnax, atque mordax cum cursu superatur, supinus dentibus, et unguibus cum canibus acriter dimicat, eius neque caro esu iniucunda est, neque damnosa pellis: adversus officientium oculorum fascinationem opi esse traditur: ob quod ex ea confectae villosae corrigiae iumentorum fraenis adduntur. Eius adipem febri correptis opitulari Sereni carmen testatur. *Nec spernendus adeps dederit quem bestia melis Sanguinem quoque siccum, et in pulvere redactum lepra infectis utilem esse perhibent.*

11. L. melis
G. μελίς
H. texon
I. broc

12. H. lepus[101]
G. λαγώς[102]
H. liebre
I. gearrfhie

Lepores pedibus leves, atque veloces (sed qui Ibernicorum canum velocitate superantur) innocua, timida carne, et pellibus utilis viventia benigna insula uberrime generat, etiam suporfoetantes quippe cui foetum alium educant, alium in utero pilis vestitutum alium implumem; alium inchoatum gerunt, qui leporis carnem saepe comedant, hos formosiores fieri, memoriae proditum est. Illis inter esculentas quadrupedes Martialis primas detulit.

Geniis

*Inter aves turdus, si quis: me iudice, certet. Inter quadrupedes gloria prima lepus.*

---

99.   Muic friain is crossed out before Aper in R. margin, muic fhian is also crossed out in L. margin. The Greek name is crossed out in R. margin and is illegible; it is replaced by ἑρραός in L. margin. After Javali in R. margin, Meles is crossed out.

100.  What appears to be 'reperit' is crossed out and 'ferunt' is inserted above the line.

101.  Gearrfhiagh is crossed out before Lepus.

102.  C. Lagos is crossed out in L. margin and inserted in Greek letters in R. margin.

Boars, or wild woodland pigs are reported to have been very fat *Boar* here; they are bellicose and fierce, when provoked they are to be feared by unarmed men.

The badger, also called Taxus, flourishes here; it is larger than *Badger* the fox: it sleeps in holes in the ground. It is pugnacious by nature and snaps when overtaken by the chase. It fights fiercely on its back with dogs, sharply with tooth and nail. Its flesh is not unpleasant to eat nor is its pelt to be despised. It is reported that it is helpful against the spell of the evil eye. Because of this, thongs of hair made from the skin are added to the bridles of baggage animals. The poem of Serenus testifies that its fat is helpful to those gripped by fever.

*Fat yielded by the badger is not to be despised.*

They say that the blood dried and rendered into powder is useful for those infected with leprosy.

*Hare*  Hares, light on the feet and fast (but who are overtaken by the speed of Irish hounds), harmless and timid animals, useful for their meat and skins, the benign island (Ireland) produces abundantly. They even conceive while still with young and bring forth one foetus, while carrying another covered in hairs in the uterus, another without hair, and another incomplete. Those who often eat the flesh of a hare, it is recorded become more beautiful. Martial referred to them as first among the edible quadrupeds.

*Martial 13.42*  *In my opinion, among the birds the thrush is a contender, if my judgement is worth anything; among quadrupeds the outstanding glory is the hare.*

*13. L. Cuniculus*[103]

*G. δασύπους*

*J. Conejo*

*I. Connin Lib. 6.*

*Roibeid Laurices*

*Martial*

*Epigrams 613, 60*

Leporum genus alterum cuniculi foecunditatis etiam innumerae, saporis gratissimi longe confertissimis agminibus hic occurunt.

Cuniculis, id est, specubus multiforibus, quos ipsi sub terra fodiunt sese abscondunt: quod et Martialem non praeterivit. *Gaudet in effossis habitare cuniculus antris.*

Eorum foetus laurices a matris vel ventre execti, vel uberibus ablati in iucundissimo cibatu habentur.

*14. L. Herinaceus*

*G. ἐχίνος*[104]

*H. Erizo*

*I Grannoig*

Herinacei degunt hic dorsum firmis aculeis defensi, ventrem innocua lanugine vestiti: super iacentia poma sese volutantes, spinis affixa in arborum, // sepiumque cavos quos habitant, portant, unum duntaxat ore tenentes. Ubi venantem sentiunt, rostro, pedibus, et omni inferiore parte contractis, in pilae formam confluunt, ne quid comprehendi possit praeter aculeos, quibus canem incaute mordentem impune vulnerant.

Mustelae, quadrupedes parvae, oblongae, tenues, animosae infestant columbaria, et alia id genus aviaria, ova exorbentes, et aves enecantes: sed mures potissimum exitio dant. Earum in Ibernia duo genera: alterae fere flavae per agros, et sylvas plerumque vagantur alterae castaneae colore in domibus etiam aberrant.

*15. L. mustela*

*G. γαλῆ*[105]

*H. Comadreja*

*I. Asoga blathnoid*

Mustelae species est viverra cui in Ibernia est magna gratia propter cuniculorum venatum nam cum sit gracilis, et oblonga in specus iniecta cuniculos eiicit: eiectique vel reti vel alia arte, ab hominibus superne capiuntur.

*16. L. Viverra*

*G. ἴκτις*[106]

*H. Uron*

*I. Fireid*

---

103. Connin is crossed out before Cunniculus. C. Dasypus is crossed out in L. margin and inserted in Greek letters in R. margin.

104. Crannoig is crossed out before Herinaceus. Echinos is deleted in L. margin and Εχίνος inserted from R. margin.

105. Lisoig is crossed out before Mustela, G. Gale is crossed out in R. margin and inserted in Greek letters from L. margin. I Lisoif is crosssed out and Asuga, Blathnoid inserted in L. margin.

106. G. Gale is crossed out and 'ικτίς inserted in L. margin. Fireid and ? Vivera are crossed out in R. margin.

*Rabbit*

*Martial Epigrams*

*L. 13, 60*

Another class of hare is rabbits, which are very fertile with a wonderful taste. They are met here in the most dense groups. They hide themselves in burrows, that is tunnels with many entrances, which they dig under the ground. This did not pass Martial by:

*The rabbit rejoices to live in hollowed out caves.*

Their young, either cut out from the mother's womb or taken away from the teats (as sucklings) are considered the most pleasant food.

*Hedgehog*

Hedgehogs live here. They are protected by firm spines on their backs, but their belly is covered with a harmless wool. They roll on fruit, lying on the ground, which become affixed to their spines. They carry them to the holes of trees and hedges which they inhabit, holding just one in their mouths. When they sense a hunter they roll themselves into a ball with their snout, feet and every underneath part contracted so that nothing can be grasped except spines, with which they wound with impunity any dog who is not careful when biting.

*Weasel*

Weasels are small quadrupeds, long, thin and lively. They attack dove-cotes and other such aviaries, breaking open the eggs and killing off the birds. But mostly they kill off mice. In Ireland, there are two types: one almost yellow, that mostly wanders through fields and woods; the other, chestnut in colour, strays even into houses.

*Ferret*

The ferret is a type of weasel greatly prized in Ireland for hunting rabbits; for since it is graceful and long, when it is put into the burrow it drives out the rabbits. When they are driven out they are captured above ground by men, either by nets or by some other device.

Martes vulpe minores, fele maiores,[107] quarum pellis hodie    *17. L. martes*[108]
in summo pretio, caro in nullo ponitur, hic frequentes. Id    *G.*
animal agile, pugnaxque canibus urgentibus, arborem celeriter    *H. Marta*
ascendit. Si nox venationem opprimit, ignis ad arboris radicem    *I. Maidri criobhoig*
accensi splendore vel stupidam vel territam feram venator in
sequentem lucem continet.

Sin diurna luce fulgente vel lapidibus, vel iaculis obruit, tunc
martes eadem, qua malum subierat, descendens, per medios
hostes, homines atque canes proeliando conatur evadere,
plerumque tamen frustra, Meminit harum Martialis.

*Venator, capta marte, superbus adest*[109]

*18. L. Spira*[110]    Martibus minores,[111] muribus maiores hic videntur sciuri,
*G. σχίουρος*[112]    colore rubidi, corpore villosi, cauda villosiores et falcati, unde
*H. hardilla*[113]    spirae nonnullis nuncupantur,[114] agilitate miri ex arbore in
*I. Cruibhoig*[115]    proximam saliunt. Nihilominus in hominum potestatem
veniunt vel in cavernis deprehensi, vel inter dissitas arbores
cursu[116] victi, vel iactu confossi. Servati in caveis festivi, mortui
pelle commendabiles: haec enim et calida vestibus addita est,
est inconsutile marsupium facit nobilibus aestimandum, si
interna tunica serica, et aureis, argenteis, vel sericis teniis, ut
in Ibernia solet, ornetur. Sciuri tempestatem praevidentes,
cavernis, qua ventus spiraturus est, obturatis, ex alia parte fores
aperire, et in hyemem pabulum providere, dicuntur.[117]

---

107.  After maoires 'hic frequentes sunt' is crossed out.
108.  Before Martes in R. margin Maoir cribhig, Martes and two illegible names are crossed out.
109.  Deleted '? las citidas mustelas ? sylvestres a Plinio dictas puto? Sylvestras semper habitant et ? more mustellarum insidiantur'.
110.  In L. margin I. Cribhoig, iera rua, sciouras and madra are crossed out.
111.  Ictis crossed out in R. margin.
112.  G. Sicurus in Latin letters is crossed out and σχίουρος inserted from R. margin.
113.  Modern Spanish ardilla.
114.  'Unde spirae nonnullis nuncopatur' is inserted from R. margin.
115.  Following Cruibhoig 6 illegible words are crossed out in L. margin.
116.  Illegible word crossed out in text after cursu.
117.  Crossed out in R. margin 'lacertae hic passim conspiciuntur sed sine venenu innocuae: nisi dormientis sub dio os fortasis ingrediatur quod vix semel vel bis toto aveo tigisse accepi'.

Martens are common here. They are smaller than foxes but larger than cats. Their skin today is very valuable, but the flesh is of no value. This animal is agile, pugnacious and, when pressed by dogs, quickly climbs a tree. If night closes in on the hunt, the hunter traps the beast until the next dawn, which is either stupefied or terrified by the brightness of a fire lit at the root of the tree. But if when daylight dawns, the hunter tries to overwhelm him with stones or javelins, then the marten comes down the same way by which he had got himself into danger, and tries to escape through the middle of the enemy men and dogs by fighting, very often, however, in vain. Martial mentions these things:

*Martial 10.37,18*     *The hunter, when he has captured the marten, proudly stands by.*

*Squirrel*

Squirrels are seen here: they are smaller than martens, but larger than rats, red in colour, with hairy bodies, the tail more hairy and sickle shaped, from which they are called *spirae*[118] by some; they are wonderful in the agility with which they jump from one tree to the next.

Nevertheless, they come into the power of men, either caught down in their warrens or beaten in the chase between widely spaced trees, or are pierced by a javelin. When kept in holes, they are amusing; when they are dead, their skin is praiseworthy. The skin is warm when added to clothes, and also it makes an unseamed purse highly esteemed by noble people, when it is adorned with silk lining inside and gold, silver or silk ribbons, as is the custom in Ireland. Squirrels predicting a storm are said to stop the nests on the side from which the wind is about to blow, and to open a door on the other side, and to store up food for the winter.

---

118.    The word means a coil, twist or tangle, referring here to the shape of the tail.

*19. L. Lutra*
*G. ἔνυδρις*[119]
*H. nutria*
*I. maidri iski*

Haud rara reperitur hic lutra: quae in terrestriumne, an aquatilium numerum referatur, id ansam dubitandi praebet, quod tam terrae, quam maris, et amnis communis sit. In mari enim, fluminibus, et mediterraneis etiam lacubus piscatur: in terra foetus parit,[120] et teneros alit. Sed quia quadrupes est utrobique agere solita, hoc loco non incongruenter ponitur. Est magnitudine vulpi suppar, colore pene castaneae, pelle non damnanda, pilis laevis cauda longa et pilosa, dentibus acutis et quibus venatori, tam homini, quam cani[121] saepe vulnus intulit, ossa // raro fregit. Cum catulus deprehensus ab homine nutritur, cicur agit, et pisces in unda captos domum saepe defert.[122]

*Vide satax apud Arist L.1 Hyst. C.1 ad Medium*

### NOXIA VIVENTIA

Hactenus animal damnosum memoravimus nullum; sunt tamen in Ibernia quatuor genera noxiorum, non quia nullomodo sint utilia, cum vel pellibus vel aliquo medicinae usu prosint, sed quia ad memorabile damnum inferendum natura sua ducuntur: ut sylvestres feles, quae non solum irritatae cum hominibus congrediuntur, sed ipsae nonnunquam solitarium provocant:

*20. L. sylv. feles*[123]
*I. fiechait*

rapax, vorax, qui boves, et oves laniat, pelle tamen laudabilis, eius carnem, quibusdam, qui commederunt, capillos cecidisse, accepi:

*22. L. lupus*[124]
*G. λύκος*
*H. lobo*
*I. macdiri*

---

119.  Enudris is crossed out and ἐνυδρίς is inserted from R. margin.
120.  Alit is crossed out before parit.
121.  'Tam homini quam cani' is inserted from the L. margin.
122.  This is later inserted under lacerta, the lizard.
123.  Cait is crossed out above 'sylv'.
124.  G. Lycos is crossed out and λύκος inserted in L. margin. Macdiri is crossed out above Lupus.

Otter

The otter is found here quite commonly. Whether it should be referred to among the number of terrestrial or of aquatic animals, is an occasion of doubt, since it is at home on the land, the sea, and the river. It fishes in the sea, in rivers, and in inland lakes. It gives birth on land and nourishes its young, but since as a quadruped it is accustomed to move about on both land and water, it is not unfittingly classed here. It is about the same size as the fox, and, in colour, almost chestnut. It has a pelt not to be despised, with fine hairs; it has a long bushy tail, and sharp teeth with which it often inflicts wounds on the hunter, be it man or dog. On rare occasions it breaks bones. When a cub is caught and is nourished by a man, it becomes tame and often *See Satax in Arist. 1.* brings back home the fish it has caught in the water. *Ch.1 towards middle*

### HARMFUL CREATURES

Up to this point we haven't recorded any animal that causes *Wild cat* injury. There are, however, four types of harmful animals in Ireland, harmful creatures not because they are completely useless (since they are of profit for their skins or for some medical applications), but because they are led by their nature to cause remarkable damage. For example, wild cats of the wood do not only fight with men when provoked, but themselves sometimes challenge a man on his own.

Wolves are rapacious and voracious, and tear to pieces cattle and *Wolf* sheep. However, their pelt is praiseworthy. As to their flesh, I have learned that certain people who have eaten it have had their hair fall out.

ut vulpes vix alia re quam pelle commendanda quae rapax, *23. L. vulpes*[125]
astuta, fraudulenta, tota in fallaces artes versa, ovibus, gallinis, *G. ἀλώπηξ*
et aliis avibus insidiatur: *H. raposa*
*I. maidri Rua*

ut mures quorum maximos sciuro paulo minores. Gallicis *24. L. mus*[126]
navibus[127] in Iberniam transportatos primum fuisse,[128] *G. μῦς*
argumentum est, quod ab Ibernis mures Franci vocantur. Hi, *H. raton*
et medii dicti Ibernici mures vestes rodunt, frumentum cibos, *I. luch*
et alias res mordent: Minimi nomine sorices sibilo prodentes,
cum cursu non valeant, facile ab homine perimuntur. Hinc
sorex suo indicio perire, is, qui se sua voce prodit, proverbialiter
dicitur. Terentius cecinit
*Egomet meo indicio miser, quasi sorex hodie perii.*

Caeterum iniuriae horum ultrices acerrimas ibidem opifex
naturae collocavit, cuas Ibernicas, sive canes relatas, quarum
ferocitate cati sylvestres, lupi, vulpes illati damni poenas capite
quam saepissime pendunt, felesque domesticas, atque mustelas
murium ausis obviam ire solitas.

---

125. Maidri rua is crossed out above 'vulpes'. Beneath vulpes 22 and lucha
are deleted The mus is crossed out followed by G. Alopix crossed out
and replaced in L. margin with ἀλώπηξ.
126. He gets the numbers wrong, it should be 23 instead of 24. G. Mys is
crossed out and μῦς inserted from L. margin.
127. From 'gallicus navibus' to 'Ibernici mures' are inserted from L. margin.
An illegible word is crossed out after 'paulo minores'.
128. Fuisse is crossed out before primum.

The fox is scarcely to be recommended for anything except for *Fox* its skin. Rapacious, cute, deceitful, and completely versed in deceptive tricks, it lies in ambush for sheep, for fowl and other birds.

The largest rodents are a little smaller than squirrels. It is argued *Mouse* that they were first transported into Ireland on French ships, because they are called French mice by the Irish.[129] These and the medium-sized ones, called Irish mice, nibble at clothes, and gnaw at corn, foodstuffs, and other things. The smallest are called shrew mice. They give themselves away by whistling. Since they are not strong in running, they are easily killed off by man. Thus, he who betrays himself by his own voice is said in the proverb to die like a shrew by its own evidence. Terence wrote:

*Ter. Eun. 5.6,22*   *Poor me, like a shrew I have perished today by my own testimony.*

However, the architect of nature[130] has placed in the same environement, as most keen avengers of the harm done by them, the Irish 'cuas',[131] or hounds, spoken of above, through whose ferocity wild cats, wolves and foxes, very often, pay the penalty with their lives for the harm they have done, and domestic cats and weasels which are wont to prevent the bold forays of rodents.

---

129.  Luch Francach in Irish means a rat. Literally, a French mouse.
130.  God.
131.  See above, page 25.

## C. XII
### Iberniae aves

Ad huius insulae volatilia calamum transferamus: quorum consideratio aquatilium relatione mihi quidem prior, atque dignior videtur: tametsi hunc ordinem Plinius inverterit. Quae enim marinarum belluarum aquilae se nobilitate aequat? Huius igitur alta generositas pluris erit aestimanda, quam balaenae humilis vastitas. Accedit avibus is cumulus dignitatis, quod Christus Redemptor noster a Joanne praenuntio Jordanis fluminis aquis tinctus spiritum sanctum sub culumbae forma latentem viderit.

## C. XIII
### Aves mansuetas terrestres[132]

Columbarum in Ibernia duo sunt genera, utrumque frequentissimum, alterum ferum nomine Palumbes, de quo inferius agetur: alterum mansuetum, & domesticum columba proprie vocatum. Columbae pavonibus amicissimae, colore variae, atque pulchrae ad solis splendorem diverse sitae diversos faciunt repercussus, attestante Lucretio:
*Qualis enim caecis poterit color esse tenebris? Lumine qui mutatur in ipso propterea quod Recta, aut obliqua percussus luce refulget. Pluma columbarum quo pacto in sole videtur: Quae sita cervices circum, collumque coronat: namque alias fit, ut clara sit rubra pyropo: Interdum quodam sensu fit, uti videatur Inter caerulem virides miscere smaragdos.*

*L. columba*[133]
*G. περιστερά*
*H. paloma*
*I. coluir colum*

Aetate quinquemestres foetificare auspicantur: coniugalis amoris symbolum, coniugii fidem minime violant: tametsi de foeminis columbi adulterii suspicionem interdum concipiant. Ante coitum haud rara[134] osculantur: a coitu quinto die ova candida, fereque bina, et anno decies, vel undecies pariunt: ea

---

132.   Terrestres is crossed out before mansuetas.
133.   G. Peristera is crossed out and περιστερά inserted from L. margin.
134.   Haud rara is inserted above the line in a different hand.

# C. XII
## Irish Birds

Let us transfer our pen to the flying creatures of this island. Consideration of these appears to me to have priority and is more worthy of reporting than of the sea creatures although Pliny[135] turned this order around. For what large creature of the sea equals the eagle in nobility? Thus, the eagle's lofty nobility is to be more esteemed than the lowly massiveness of the whale. To the birds goes this accumulation of dignity, because Christ our redeemer, when baptised by John, his forerunner, when he had been dipped in the waters of the River Jordan, saw the Holy Spirit hidden in the form of a dove.

# C. XIII
## The Tame Land Birds

In Ireland, there are two types of dove, both very common, *Dove* one wild, called the wood pigeon, concerning which mention will be made below, the other – the tame, domestic type – is characteristically called a dove. Doves are very friendly to peacocks, variable in colour and beautiful; when placed in different positions towards the sunshine they produce different reflections as attested by Lucretius:[136]

*For what colour can there be in blind darkness? Nay a colour is changed by the light itself, according as it reflects back, struck by direct or oblique light; even as the dove's plumage shows itself in the sun, lying about the nape and encircling the neck; for at times it is red as a clear carbuncle, again view it in a certain way and it comes to appear a fusion of emerald green with blue.*

They begin to breed at five months of age. They do not violate the bond of marriage, a model of conjugal love: however, as regards the hens the males sometimes conceive the suspicion of adultery. Before intercourse, they often kiss. As a result of intercourse they lay white eggs on the fifth day, two at a time

---

135.   Pliny devotes his Natural History III, Book 9, to sea creatures and Book 10 to birds.
136.   Lucretius *De Rerum Nat.* II 798–804.

interdiu mas, noctu foemina incubant: vicesimo incubatus die pullorum par, marem prius, inde foeminam saepius excludunt. Si maris copia non sit, foeminae ova quidem dicta Graecis hypenemia, & Zephyria; sed aliis minora, humidiora, sapore iniucundiora, irritaque, et sterilia (quibus avis haudquaquam gignitur) favonii venti flatu, et veneris imaginatione concepta edere feruntur. Octonos annos plerumque vivunt.

A columbis vero quia sumpsimus exordium alias domesticas aves, ac prius terrestres, id est in terra duntaxat degentes, et digitos natatui ineptos habentes addamus: postea de feris dicturi, ut a notis ad ignotiora procedat oratio.

*L. pavo*
*G. παώς*[137]
*H. Pavon*[138]
*I. Piacoig*

Hic columbarum amicus pavo domesticus est vario pennarum fulgore, conchataque cauda quam pulcherrimus, Lucretii versibus describitur.

*Caudaque pavonis larga cum luce repleta est: Consimili mutat ratione observa colores. Quin quoniam quodam gignuntur luminis citu, Scire licet, sine eo fieri non posse, putandum.*

Ab aetatis anno tertio radiantes, atque gemmantes colores incipit fundere, laudatusque maxime expandit, Ovidio teste:

*Laudatas*[139] *homini volucris Iunonia pennas*
*Explicat, et forma multa superbit avis*

Omnibus annis amissa cauda, donec alia renascatur, latibulum pudibundus, atque moestus quaerit: in annum trigesimum vitam saepe perducere traditur. Foeminae a trimatu pariunt ova, anno primo singula, vel bina, sequente quaterna, vel quina caeteris duodena non amplius: et vel noctu, vel in latebris, vel in celso loco incubantes, ne a maribus incubantium desiderio frangantur ova, partus dies circiter viginti septem excludunt.

---

137. G. Paos is crossed out and παώς inserted. This, however is incorrect as the correct name for Peacock is ταώς.
138. Modern Spanish pavón.
139. The MS reads laudatus but it should be laudatas agreeing with Pennas.

and they produce ten or eleven times a year. The male incubates during the day and the female at night. They hatch a pair of chicks on the twentieth day, more often the male first then the female. If there should not be a sufficient supply of male birds, the females are said to produce eggs still, the ones called by the Greeks 'wind eggs' and 'Zephyr',[140] smaller than others and more moist, and more unpleasant to the taste, and useless and sterile – birds are never born form them – conceived from a blast of the west wind and the fantasy of love. Most doves live eight years.

But since indeed we have started with doves, let us add on other domestic birds, and first the land birds, that is those who live only on land, and let us add those with toes useless for swimming. After this, I will speak about the wild birds so that the tale may proceed from what is known to the less known.

*eacock*

Here, the friend of the doves is the domestic peacock, which is a very beautiful sight with its variegated feathers and shell-shaped tail. It is described in the verses of Lucretius:[141]

*And the peacock's tail when it is suffused with plenteous light, in like manner changes the colours as it turns: and since these colours are caused by a certain impact of light you may be sure that the change must not be thought possible without it.*

From his third year of age, he begins to pour forth radiant and jewel-like colours and when praised spreads out [his tail] to his greatest extent, as Ovid testifies:[142]

*The winged creature of Juno unfolds its feathers when they are praised by men. And the bird is very proud of its beauty.*

It loses its tail every year and, ashamed and sad, it seeks a hiding place until another grows. It is reported to live often to thirty years. The females lay eggs from three years of age, in the first year a single or two eggs; in the next, four or five at a time; in the rest, not more than twelve at a time. They sit on the eggs either at night or in hiding places or in a raised up place to avoid their being broken by the males through desire for the sitting hens. They hatch their young on about the twenty-seventh day.

---

140.  Zephyr is the West Wind, so 'west-wind eggs'.
141.  Lucretius *De Rerum Nat.* II. 806–809. In this passage Lucretius is propounding his theory of light and that colour depends on light.
142.  *Medicamina femineae faciei*, 33–4.

Gallinarum aliae ferae suo loco considerandae, aliae mansuetae, *L. Gallina*
ut Villaticae, et Africae. Villaticae sunt illae vulgatissimae, quas *G. ἀλεκτορίς*[143]
non modo in villis, sed etiam in oppidis nasci videmus: & carne, *H. Gallina*[144]
pluma, ovis (modo ab urinis abstineatur) hominibus magno *I. Kearc*
usui esse fatemur:aves foecundissimae; sed quae, ovo edito,
inhorrescere, seseque excutere, traduntur. Earum nobilitas
erecta crista non parum spectatur.

Mas quoque gallus gallinaceus elata crista, ardua cervice, *L. Gallinaceus*
fastuoso incessu, falcata cauda ferocitatem prae se fert, coelum *G. ἀλεκτρυών*[145]
saepe aspiciens. Regnum sui generis, in quacumque domo *H. Gallo*
sit, affectat: de quo inter gallos acerrime dimicatur. Victus vel *I. Coileach*
dimicatione moritur vel silens servitium patitur: victor regnat.
Tamquam hominum vigil, media nocte, altissimo sole, albente
coelo, alis, cantuque maxime obstrepit. Virg.
*Excubitorque diem cantu patefecerit ales.*

*L. Meleagris Guttata*[146] Meleagrides Africae Gallinae nuncupatae, quod ex Africa
*G. μελεαγρίς* in Italiam primum translatae credantur, Ibernis Iudaicae, et
*H. Gallina Morer* Gallicae nominatae, grandiores villaticis variae, gibberae ova
*I. Kearc iulach* punctis distincta pariunt.

# C. XIV
## Aves mansuetae aquaticae

*L. Anser* Ad mansuetas palmipedes, et aquaticas accedamus. Anseres
*G. χήν*[147] Ibernia, et feros, et domesticos alit. De illis postea dicetur. Hi
*H. Ganzo*[148]
*I. Geae*

---

143. G. Alectris is crossed out and ἀλεκτορίς inserted from the L. margin.
144. Modern Spanish Gallina de Guinea.
145. G. Alektryon is crossed out and ἀλεκτρυών inserted from the L. margin.
146. G. Merleagris is crossed out and μελεαγρίς is inserted from R. margin (μελεαγρίς is crossed out in R. margin above μελεαγρίς).
147. G. Chen is crossed out and χὴν put next to it.
148. Modern Spanish Ganso.

Of the hen family some are wild and need to be considered *Hen* in their own place, others are tame such as the farmyard, and the Africans. The farmyard hens are those very common ones we see produced not only in villages but also in towns, and we admit that in their flesh, feathers, and eggs, provided one avoids their urine, they are of great use to man: the birds are most fertile, but they are reported to bristle and shake themselves when they lay an egg. Their nobility can be seen not in small measure when their comb is erect.

The male also, the dunghill-cock, carries his ferocity before *Cock* him, with his comb lifted up, his neck erect, his haughty gait, and his sickle-shaped tail, while he looks often to heaven. He strives after the kingship of his own kind in whatever house he may happen to be: for this reason, fighting is fierce among cocks. The conquered either dies in battle or suffers servitude in silence. The victor reigns. As though the watchman of men, in the middle of the night, or the middle of the day, or at daybreak, he makes a great clamour with his wings and his crowing. Virgil:[149]
*The sentinel bird has opened up the day with crowing.*

*Guinea hen*  Guinea hens, called African hens because they are believed to have first been brought to Italy from Africa, are called Jewish and French hens by the Irish. They are larger than the farmyard hens, of varying colours, hump-backed, and they lay eggs marked with distinctive spots.

## C. XIV
### Tame Aquatic Birds

*Goose*  Let us go to the tame, broad-footed, and aquatic birds. Ireland produces both wild and domestic geese. I will discuss the former later. The latter copulate in water and for the most part, lay their eggs in spring or summer; they are not least commendable for their sleep-inducing down.

---

149.    Virgil Moretum line 2. N.B. This is in the *Appendix Vergiliana*.

in aqua coeunt, vel vere, vel aestate plerumque pariunt; plumis etiam somniolentis non mediocriter commendabiles: quasi nocturnas excubias agentes adveniente aliquo suas vigilias clangore testantur. Ovid.

*Humanum longe praesentit odorem*
*Romulidarum arcis servator candidus anser*

Ansere minores anates ab assiduitate natandi nomen apud Latinos adeptae in Ibernia alterae quoque ferae, de quibus postea loquemur, alterae mansuetae. Ex aqua, et fovea etiam in sublime prorsus se tollunt, quassis pennis, unde quas sigipennae Varroni appellantur: tota carne esu non ingratae; sed pectore, cerviceque potissimum Martiali laudatae.

*Tota sibi ponatur anas, sed pectore tantum*
*Et cervice sapit: caetera redde coquo*

In earum ovis non raro duo vitelli reperiuntur. Ad diem usque Divi Patricii Iberniae praesidis nidos plumeis, vel herbeis operculis tegunt.

Gignit haec insula cygnos, vel olores, et feros, et mansuetos, ansere maiores, colore candidos, cervice praelongos: Ovidis carmine pulcherrime descriptos:

*. . . collumque a pectore longe*
*Porrigitur; digitosque ligat iunctura rubentes*
*Penna latus velat: tenet os sine acumine rostrum*

Aquilam aversantur, mortem sibi appropinquantem cantu celebrare, a Martiale referuntur.

---

150.   G. Nessa is crossed out and νῆσσα inserted from the R. margin.
151.   Modern Spanish Ánade.
152.   G. Cygnos is crossed out and replaced by κύγνος from L. margin.
153.   The correct Greek word for swan is κύκνος.
154.   Modern Spanish Cisne.

As if maintaining nightly guards, when anyone approaches they announce their alertness by a loud noise. Ovid.[155]
*He scents from afar the smell of humans,*
*the saviour of the citadel of Romulus' descendants, the white goose.*

Duck

Ducks, which are smaller than geese, acquired their name among the Latins from being continually involved in swimming.[156] In Ireland there are two types, one wild (also concerning which we will speak later), the other tame. They can rise straight up into the air out of water and even out of a pit, shaking their feathers, because of which they are called feather-shakers by Varro. The flesh as a whole is not unpleasant to eat but the breast and neck are most praised by Martial.

*Martial Epigrams*
3, 52

'*Let the whole duck be placed before you but it only has a good taste in the breast and neck; give the rest back to the cook.*'

Swan

In their eggs two yolks are often found. Up to the day of St Patrick, the Protector of Ireland, they cover their nests with blankets of feathers or grass.

This island produces *cygnos* or *olores* (swans) both wild and domestic. They are bigger than geese, white in colour, with very long necks. They are described most beautifully in the poem of Ovid:[157]
*His neck stretches out a long way from his chest,*
*and a web binds his red toes. A wing covers his flanks,*
*and a beak without a point takes over his mouth.*

They avoid the eagle and are reported by Martial to celebrate their approaching death with a song:[158]

---

155. O'Sullivan Beare says Ovid, but the quotation is from Lucretius, *De Rerum Natura* 4, 682–3.
156. Anas-Anatis – a duck from natare, to swim.
157. *Metamorphoses* 2, 374–6. Here O'Sullivan Beare is borrowing Ovid's account of Cycnus' metamorphosis into a swan, rather than quoting a description of an actual swan.
158. Martial xiii. 77.

*Dulcia defecta modulatur carmina lingua*
*Cantator, cygnus, funeris ipse sui*

Quod tamen falsis experimentis fuisse collectum, Plinius arbitratur. Equidem legi, caput cygni spectantis ad mortem penna introrsus nascente transfigi, et ob id ab ipso flebilem vocem emitti, dolore cogente. Has in Ibernia mansuetas aves reperio.

## C. XV
### Aquilae species

Caeterae, nisi cicures fiant, ferae sunt: quarum etiam aliquas memorare, ac prius terrestres, sive digitos habentes, opere pretium puto, si ab aquila omnium quas cognoscimus regina fortissima, initium sumatur. Ibernica nobilissima est in arboribus, atque petris nidificans; ova terna parit, quorum unum taedio nutrendi expellit: ex aliis duobus pullos trigesimo circiter die incubationis excludit: fame, quia superius rostrum in tantum excresit, ut aduncitas aperiri nequeat, oppetit. Aquilarum genera sex numerantur. Omnium maxima gnesios colore subrutilo inficitur.

*L. Aquila*
*G. ἀετός*[159]
*H. Aquila*[160]
*I. Fiuluir*

*Gnesios Plangus*

Plangus multis modis, plancus, morphnos, Porcnos, Naevia, Antaria dicitur, aquilarum nigerrima, testacea esculenta rapta e sublimi in lapidem iaciendo, frangit: circa lacus degens natantem avem petit: impetita sese aqua margens ad arundinis condensae, vel littoris impediti perfugia tendit. Plangus in latus sese deiiciens, nanti suam sub aqua umbram a littore ostendit: qua pavida avis in diversa saepe emergit, donec sopita et lassata rapiatur.

*Pigagrus Baull iar*
*Tieri., Valeria*

Gnesio pygargus Magnitudine proximus est, cauda albicans, in campis, et oppidis mansitans.

---

159.  G. Aetos is crossed out in R. margin and ἀετός inserted from L. margin.
160.  Modern Spanish Áquila.

*The swan sings sweet songs with a failing tongue,*
*himself the poet of his own death.*

This fact, however, Pliny considers to have been acquired by
false proofs. Indeed, I have read that the head of a swan looking
at death is transfixed by a feather growing from within, and that
it is because of that it lets forth from itself a doleful cry, driven
on by sorrow. I find these tame birds in Ireland.

## C. XV
### Types of Eagle

The other birds, unless they are tamed, are wild; some of them I   *Eagle*
consider it worthwhile to mention, and first the terrestrial ones
with talons, if we make a beginning with the eagle which is the
bravest queen of all the birds we know. The Irish one is the most
noble, and nests in trees and among rocks. It lays three eggs at a
time of which it throws out one because of difficulty in feeding.
From the other two, about the thirtieth day of incubation it
hatches out two chicks. It dies through hunger because the upper
beak grows out so much that the curvature prevents its being
opened. The greatest of all these, the native one,[161] is tinged
*Golden eagle*    with a reddish colour. The Plangus is called by many names, the
*Plangus*    Plancus, Morphnos, Porenos, Naevia and Antaria. The blackest
of eagles, it snatches edible creatures with a shell, and, throwing
them from on high on to a stone, he breaks them open. Dwelling
around lakes it attacks swimming birds. The bird which is
attacked makes for the refuge of thick reeds or overgrown shore.
The Plancus, diving itself on to the flank shows its own shadow
to the swimming bird under the water from the shore. In this way
the frightened bird often tries to escape in a different direction
until, stunned and worn out, it is captured.

*Hen-harrier*    The white-tailed eagle is next in size, after the gnesios. It has a
*White-tailed*    white tail and frequents plains and towns.

---

161.   The Golden Eagle.

Valeria nominibus aliis, fulva, Pulla, Leporaria, melenaetos vocata // corpore minima, viribus praecipua, colore nigricans, in montibus commorari solita, sine clangore, et murmuratione agit, suosque foetus alit, cum aliae aquilae pullos fugent. Ab omnibus, quas recensuimus, aetitem lapidem, ova refrigerandi causa, in nidis collocari, memoriae proditum est.

Pygargus, et Valeria non minores tantum quadrupedes venantur, sed etiam cum cervis proelia committunt: quorum cornibus insidentes, pulverem volatu collectum in oculos excutiunt, ora pennis verberando, donec in rupes praecipitent, ut alto casu occisorum carnibus vescantur.

Eadem etiam arte cervos aggrediuntur ossifragae ab ossibus *Ossifraga*
frangendis,[162] et barbatae a plumis e mento pendentibus duplex nomen sortitae: pygargis non multo minores: anguillis, et aliis piscibus vivunt: non modo suos; sed etiam ab aliis aquilis eiectos foetus nutriunt.

Ex aquilarum diverso coitu nasci haliaeetus fertur, quae suos *Halietos*
pullos etiam num implumes percutiendo radios solis adversos *Iaskaire Keanainn*
intueri, cogit: et conniventem sive lumina flectentem, sive flentem, velut adulterinum, et degenerem e nido praecipitat, ossifragae educandum firmiter vero lumen aspicientem alit: ipsaque perspicacissima pollens oculorum acie, in piscem sub aqua visum ita liberato [ex alto][163] corpore ruit, ut undis pectore discussis illum capiat: ad id instructa altero pede palmipede nando aptissimo, altero adunco rapiendo accomodato; sed apprehensum pondus cum trahere ob gravitatem non valet, et ob aviditatem dimittere non vult, sub unda respirare, non potens, nonnumquam mergitur.

---

162.   After ossibus an illegible word is crossed out. After soritate 'ian fuinn' is crossed out in R. margin. After minores 'anni'?
163.   Ex alto is omitted by O'Donnell.

The Valeria is called by other names tawny,[164] dark, the hare-eagle, or the black eagle. It is very small in body, but of outstanding strength, blackish in colour, and tends to stay on its own in the mountains, proceeding without making a loud noise and screaming. It nourishes as its own, the young that other eagles drive away. It is recorded by everybody we have consulted that the eagle stone is placed in the nests for the purpose of cooling the eggs.

The Pigargus[165] and the Valeria[166] not only hunt the smaller quadrupeds, but even join battle with stags. They settle on their horns and shake dust collected in flight into their eyes, striking their faces with their wings until they force them headlong over the cliffs so that, when they are killed by the high fall, they may feed on their flesh.

By the same manoeuvre, Osprey also attack stags. They get *Osprey*[167] their name *Ossifraga* from breaking bones with their beaks, and a second name *Barbata*[168] from feathers hanging from their chins. They are not much smaller than the white-tailed eagles. They live on eels and other fish. They nourish not only their own but the young thrown out by other eagles.

The osprey is said to be born from the diverse sexual union of *Osprey* eagles. The female compels her own chicks, by striking them, even while still unfledged, to gaze directly at the rays of the sun, and casts down from the nest any one of them that blinks or turns its eyes away or weeps, as if not her own and unworthy of the race, leaving it to be reared by the Osprey. But she rears the offspring that looks firmly at the light. With the power of this very acute vision she can attack a fish seen under the water, with such a thrust of the body from on high that, cutting the waves with her chest, she captures it. She is equipped for this with one webbed foot most suitable for swimming, while the other has talons designed for snatching. But when she is not strong enough to lift the weight captured and she does not wish to let

---

164. Virg. *Aeneid* 11, 751.
165. White rump type of eagle.
166. Type of eagle also called Melanaetos. Pliny 10,3,3,§6.
167. Pliny 30,7,30 §6.
168. Bearded.

Aliud aquilarum genus addunt nonnulli percnopterum, vel       *L. Perenopterus*
Oripelargum, vulterina specie, minimis alis, magna corporis    *G.*
mole, ieiuna aviditate, querula mur murmuratione, degenerem,   *H. Alhorracho*
et imbellem, quippe solitam a corvu etiam verberari: (quam in  *I. Ian Fionn*
Ibernia esse non dum habeo compertum).[169]

## C. XVI
### Accipitris species

Cum aquila bellum gerit avis alia corpore quidem, viribusque    *Acciptor*
minor, animo tamen prope nobilior (usque adeo intestinum,[170]  *G. ἱέραξ*[171]
ut ambae cohaerentes, inque terram dilapsae aliquando           *H. Azor*
deprehendantur) accipiter ab avibus male capiendis nomen        *I. Seabhac*
habens in id adunco rostro falcatisque (unde falco dicitur)
unguibus armata: inter aves magnitudine mediocris: vicenis
diebus ex ovis pullos educens, et in Ibernia frequens.

Non vero omnis accipiter cum Aquila audet concurrere;
sed nobilissimus duntaxat: cuius in Ibernia duo sunt
genera: maritimus in rupibus, sylvester in sylvis nidificans et
commorans. Uterque sublimipeta est: nam avem in sublimi, ut
arbore, vel turre sedentem, vel circa arbores cursitantem, vel in
aperto volantem capit. De quo Virgilius est intelligendus

*Aeneid II*

*Quam facile accipiter saxo sacer ales ab alto*
*Consequitur pennis sublimem in nubo columbam,*
*Compressamque tenet, pedibusque eviscerat uncis.*

---

169.   Quam in Ibernia esse nondum habeo compertum is crossed out with
       a single line.
170.   Inertnecinum is crossed out and replaced by intestinum.
171.   G. Hierax is crossed out in R. margin and G. ἱέραξ inserted from L.
       margin. I. Seave is crossed out in R. margin and Seabhac inserted
       underneath it.

it go because of greed, not being able to breathe under water, she sometimes drowns.

*Kite*

Some people add on another family of eagles, the percnopterus[172] or oripelargus. It resembles the vulture, has very small wings, but a large body, a huge appetite and a plaintive cry. It is a species unworthy of the race, not inclined to fight, in as much as it even tends to be beaten by a crow. (I have not yet discovered that the species exists in Ireland.)

# C. XVI
## Types of Hawk

Another bird, the hawk, wages war with the eagle, smaller in *Hawk* body and strength, but almost nobler in spirit. Their conflict is so desperate that the two are sometimes found sticking together after falling to earth. The hawk gets its name from the evil habit of catching birds.[173] For this it is armed with a hooked beak and sickle-shaped talons (from which it is called the falcon).[174] It is middle sized among birds, hatches out its chicks in twenty days, and is common in Ireland.

However, not every hawk dares to fight with the eagle, but only the most noble one. There are two types in Ireland, the maritime one which nests and abides in cliffs [peregrine], and the woodland type which nests and abides in the woods [goshawk]. Both seek the heights, for they capture birds on high, such as on trees, or sitting on towers, or wandering around trees or flying in the open. Concerning this one should see Virgil:

*Aeneid XI 721–723*   *As easily as the sacred hawk flies from the high rock,*
*overtakes on his wings the dove high up in the cloud,*
*holds it crushed, and eviscerates it with its crooked claws.*

---

172. Dusky winged. 'Oripelargus' is a mountain stork. O'Sullivan is uncertain of the terminology. Éan Fionn is a kite in Irish.
173. This draws on the connection between the word 'accipiter' and the verb 'capio', to catch.
174. Falco from falx sickle (sickle-shaped claws).

Utriusque pullus ab hominibus educatus cicur, et ad id aucupium laudatissimus fit annuente Martiale.

*Praedo fuit volucrum, famulus nunc aucupis idem:*
*Decipit, et captas non sibi meret aves.*

*Astur*
*Mearluin*[175]
*L. Nisus*
*G. Νίσος*[176]

Hic et aliae sunt accipitris duae species: quibus homines ad aucupium utuntur; sed ignobiliores, ut astur, et nisus; quorum hic humipeta e terra tollit aves.

*H. gavilan*[177]
*esmerjon*[178] *alcotan*
*I. speoroig*
*L. milvus, milvius*
*G. ἰκτῖνος*[179]
*H. milano*
*I. priachain na ghearc.*
*Pucaire Goithi*

Alios accipitres homines huius insulae ad aucupium non alunt: cuiusmodi est milvus, qui rapax, famelicus, querulus gallinas aggredi non ausus, earum pullos rapit: quibus insidiatur in aere suspensus, et expassas alas intervallo raro quatiens. Inter volandum crebris caudae flexibus instar navis sese gubernat. Sed dum ipse fur anumalcula imbecilla circumvenit, a sublimipeta praedone saepe appetitur: contra quem tamen se non nunquam defendit, supinus unguibus, rostroque pugnando.

Silentio praetereundus non est tinnunculus:[180] qui aspectu, voceque alios accipitres fugando, columbas, quibus amicissimus est, incolumes reddit. Corpore milvo multo minor, niso suppar, rostro brevis, aduncus, pullus, oculis niger, cauda longus, pedibus pene pallidus, nigris lineis plumas plerumque lividas distinguentibus. Avium falcatarum solus supra quaterna ova edit.

*L. Tinnunculus*
*G. κενχρίς*[181]
*H. cernicalo*[182]
*I. ruan fhaili*

175. Nisus in L. Νίσος is Greek. A King of Megara father of Scylla who to gain the love of Nisus cut off her father's purple hair on which the safety of his Kingdom depended, whereupon Nisus was changed into a sparrow hawk and Scylla into the bird Iris, Verg. G.1. 404.
176. G. Nisos is crossed out in L. margin and νῖσος inserted from R. margin. The first letter of νῖσος looks more like λ than ν but λῖσος does not exist.
177. Modern Spanish gavilán.
178. Modern Spanish esmerejón.
179. G. Hictinos is crossed out in L. margin and ἰκτῖνος inserted from R. margin. G. Ixtin is crossed out above ἰκτῖνος in R. margin.
180. Graecis cenchris is crossed out.
181. G. κενχρίς I could not find. In Lewis, Short Latin Dictionary under Cenchris – κερχρίς a 'kind of hawk'. However, κεγχρίς in Greek Dictionary = Millet seed.
182. Modern Spanish cernícalo.

*Martial 2, 216*     The chick of both hawks when brought up by men becomes tame and very praiseworthy for this type of fowling. As Martial notes:
*Once he was a plunderer of birds, but now, as the slave of the hunter, he catches them and laments that the birds have not been caught for himself.*

*Sparrow-hawk*     Here there are two other species of hawk which men use for fowling, but they are less noble, namely the merlin and the sparrow hawk. The latter is low-flying and snatches the birds from the ground.

*Kite*     Other hawks the men of this island do not rear for fowling. Of this type is the kite who is grasping, famished, querulous and does not dare to attack hens but snatches their chicks; these he attacks having been suspended in the air flapping his outstretched wings at rare intervals. During flight they steer like a ship by frequently bending their tail. But while the thief *Another kite* himself is attacking the weak little animals, very often he is attacked by a high-flying predator, against whom however he sometimes defends himself on his back by fighting with his claws and his beak.

The kestrel must not be passed over in silence.[183] By putting the *Kestrel* other hawks to flight with his appearance and voice he protects the doves with whom he is most friendly. He is much smaller than the kite, nearly equal in size to the sparrow hawk, he has a short beak, which is hooked. He is dark coloured, with black eyes, a long tail, almost pale feet, and with black lines marking the feathers, which for the most part are blueish in colour. It is the only one of the sickle clawed birds who lays over four eggs at a time.

---

183.   Pl. 10,52,109.

*L. Buteo*[184]
*G. τριόρχης*
*H.*
*I.*

Tinnunculo maior, milvo aequalis, et socius buteo: quod tres testiculos habet: triorches Graecis nominatus, non fuit omnino silentio involuendus.

*L. Cuculus*
*G. κόκκυξ*[186]
*H. Cuccilo*
*I. cuach*

Quibus adnumerandus haud quaquam fuisset cuculus,[185] sive coccyx, cum // aduncis unguibus non[187] instituatur nisi et eum ex accipitrum genere esse, scriptores tradant (tradidissent?):[188] inter quos tamen de eius natura explicanda minime convenit,[189] cum alii esse accipitrem, qui externam duntaxat figuram certo anni tempore mutat, affirmant, eo argumento adducti, quod cum cuculus apparet, accipiter, cui similis est,[190] non conspiciatur: alii[191] ex Asturis, sive nisi, columbaeque coitu gigni, sunt authores. Id certe constat, illum colore niso similem esse, ab accipitribus interimi, ova plerumque singula, raro bina, et semper in nidis aliarum avium, maxime currucarum, a quibus eius pulli educuntur, parere: vere procedere, caniculae

*L. Curruca*
*G. ἐπιλαίς*[192]
*H. curruca*
*I. cuacan coachain*
*caucan caelog riach*[194]

ortu occultari: in Ibernia vix aut ne vix quidem sine curruca comitante videris:[193] vulgoque receptum esse quotannis die Divi Patricii Martii mensis decimo septimo Ardmachae in spino sedere, et oriente sole primum canere. Non es quid alias accipitris species longius insectemur.

---

184.   G. Triorches crossed out in L. margin and τριόρχις inserted from R. margin.
185.   Haud quaquam is crossed out and reinstated in space above.
186.   G. Coccyx crossed out and κόκκυξ inserted from the R. margin.
187.   Non and nis are inserted above the line for two illegible words.
188.   Tradant is crossed out and ? tradidissent inserted above line in a different hand.
189.   Eum crossed out.
190.   Inserted above the line, qui? crossed out.
191.   Alii crossed out before alii.
192.   G. Hepilais crossed out in R. margin and ἐπιλαίς inserted from L. margin.
193.   Videri inserted above the line – should be videris.
194.   Coachain and finnin feor are crossed out before the Spanish and again after the Irish.

*Buzzard*

Larger than the sparrow hawk, the buzzard[195] is equal in size to and a companion of the kite. Because it has three testicles and is called triorches by the Greeks, it was not to be passed over altogether in silence.

*Cuckoo*

The cuckoo, or coccyx, should in no way have been counted among these since it does not have sickle-shaped claws, had writers not handed down the notion that it too is from the family of hawks. Among them, however, there is no agreement about how to explain its nature, since some authors assert that it is a hawk which merely changes its external appearance at a certain time of year, influenced by the argument, that when the cuckoo appears, the hawk to whom it is similar, is not to be caught sight of. Others write that it is born of the copulation of the merlin, or the sparrow hawk, with a dove. There is agreement for sure that it is similar in colour to the sparrow hawk, that it is killed by hawks, that it lays one egg, rarely two, and always in the nests of other birds, mostly of the hedge sparrow by whom its chicks are reared, that it emerges in the spring and disappears at the rising of the dog star, that in Ireland scarcely or never will you see it without a hedge sparrow accompanying it, and that it is *Hedge sparrow* popularly believed that every year on St Patrick's Day, on the 17th day of March, it sits on the spire at Armagh and first sings at the rising of the sun. There is no reason we should spend any more time on the other species of hawk.

---

195.   A kind of hawk. Pl. 10,8,9.

# C. XVII
## Aves nocturnae

*Coileach ithi*
*Keabhann cait*
*Fulchadan*
*Olchoochan Skreach*
*rheilgi*

Aves nocturnas addamus nam ad hunc locum partinent, aliquae aduncis[197] unguibus muniantur ut bubo, aluco, ulula noctua, asio,[198] ciconia. Bubo non minor quam aquila, gallinaceo maior Aluco, compar ulula, buboni similis noctua, ea minor Asio.

*L. Bubo[196]*
*G. βυάς*
*Keabhainn cait*
*Coilleach ithi screach*
*reiligi*

*Arist. Hist. 1.8., Hyst*
*c.3. L.9 c, 28–34*

Principio Bubo non desst hic, quia bove cuius mugitum imitari voce videtur, dicitur:[199] non cantu, sed fletu vocalis: unde cecinit Virgilius
*Seraque culminibus ferali carmine bubo*
*Visa queri, et longas in fletu ducere voces*

Cui Philomelae author assentitur:
*Bubulat horrendus ferali carmine bubo*

*ὠτίς Asio I. Auritus*
*Vide Plin. L. 10 23*

*L. Ulula,[200] Asio*
*G. Ἀιγώλιος*
*H. Autillo*

Bubone minor ulula, ab ululando quia moestum querulumque sonum edit, nuncupata Latinis quoque Asio, et Graecis aegolios dicta plumeas aures eminentes habet.

---

196. There is a lot of crossing out in R margin and he uses an insertion mark o-o to insert the Irish names in the L. margin. G. Buas, followed by two illegible word, is crossed out in R margin and βυάς is inserted from L. margin. H. Bubo crossed out and reinstated in L. margin. Three illegible words crossed out, underneath Keabhainn Cait, Coilleach Ithi, then Skreach Reilige is crossed out and reinstated underneath. Insertion mark text deleted: Coileacan ithi, keabhann cait, tulchadan, olchouchan, skreach reilige are in the left margin.
197. After nocturnas 'quae cum uncis uncuibus falcatibus rostro muniantuir, sed hunc locum pertinent, adamus' is crossed out and the above text from 'addamus to aduncus' inserted in a different hand above the line.
198. From muniantur to asio is inserted from L. margin.
199. Above the line inserted in a different hand is an illegible line.
200. G. Aegolios is crossed out in L. margin and Ἀιγώλιος inserted from R. margin. I. Gahbai roith crossed out in L. margin and not replaced.

# C. XVII
## Nocturnal Birds

*Owl*

*Long-eared owl*

*Barn owl*

Let us add nocturnal birds, for some are appropriate in this *Owl* place and some protect themselves with hooked talons, such as the eagle owl,[201] the Aluco, the screech-owl, the night owl,[202] the horned owl, and the Stork.[203] The eagle owl is not smaller than the eagle, the Aluco is smaller than a cockerel, the screech-owl is equal in size to one. The night owl is similar to the eagle owl and the horned owl, smaller than the night owl.

*Arist. Hist. Book. 8*
*hist. c. 3. Book. 9 c.*
*28 & 34*
*Virg. Aen. IV, 462–3.*
*ὗτος*

First of all, the eagle owl is mentioned here because it is named from the bull[204] whose bellowing it appears to imitate with its voice, crying not in song, but in mourning, as Vergil wrote: *Late upon the roof tops with funereal song the eagle owl is seen complaining and extending its long cries in a wail.*[205]

The author of Philomela agrees with Vergil.          *Eagle owl*
*The fearful eagle owl screeches with funereal song.*[206]

*Virg. Aen. 8,55.*
*Long-eared owl*
*Pl. 10,33,66*

The long eared owl (ulula)[207] is smaller than the eagle owl, so named from its wailing because it emits a sad, querulous sound. The ulula, also called Asio (long-eared owl) by the Latins and Aegolios by the Greeks, has long, feathered ears.

---

201.  Pl. 10,16,34.
202.  Pl. 10,19,39.
203.  Ciconia, a stork is an obvious error. O'Sullivan Beare discusses this bird on p. 55.
204.  Βοῦς is Greek for a bull. O'Sullivan Beare seems to think that βυὰς Eagle owl and Βοῦς a cow are related words. Bubo in Latin is given as an owl, horned owl.
205.  'Solaque culminibus ferali carmine bubo / saepe queri et longas in fletus ducere voces.' O'Sullivan Beare's version is different from what is now the standard text.
206.  Auct. Carm Phil 762 37 *Anthologia Latina* p. 248.
207.  Aluco, noctua, ulula are described as types of owl. The owls presently in Ireland are (1) Tyto alba: barn owl; scréach reilige; (2) Asio otus: long-eared owl; ceann cait; (3) Asio flammeus: short-eared owl; ulchabhán réisc (a winter visitor).

*L. Noctua*
*G. γλαύξ*[208]
*H. Lechuza*
*I. Tuluchadain*
*Collocat hic caput de vespertilione et pico martio*

Ei haud par, gallinae fere aequalis magnitudine, noctua, ampla capite, a multis avibus circundata supina acriter dimicat, ab aquila non nunquam adiuta: cornicem praecipuo odio prosequitur: a nocte non incongruenter nomen adepta noctua et nicticorax:[209] Nam quamvis glaucos, atque pulchros oculos habeat, interdiu tamen est acie videndi hebete, vitio[210] caeteris etiam nocturnis avibus, quas recensuimus communi: ob quod die latentes, nocte venatum egrediuntur.

*L. Vespertilio*
*G. νυκτερίς*[211]
*H. Murcequillo*[213]
*I. Fioltholg*
*fioltiog leathair luch eair (ialtog, mioltóg leathair)*

Quae causa nos ab eis in vespertilionis reputationem ducit. Hic enim[212] solis tamen splendorem debilibus oculis haudquaquam sustinet: noctu mane, vespere alis pelliceis, sive membraneis volet. Hinc ei a vespere nomen inditum fuisse Ovidius est author. *Lucemque perosae Nocte volant, seroque tenent a vespere nomen.* Aures, dentes, ubera, murisque caput, colorem, pillos habet: pullos vivos parit lacte nutrit: volitans amplectitur. Unde fuit inter authores[214] naturam rerum scrutantes id dubitatum, avibusne an terrestribus esset adscribendus. (Dic hic de pico martio. Inserted from the left margin.)

---

208. G. Nyctiacora is crossed out in L. margin and G. γλαύξ inserted from R. margin.
209. Noctua and Nicticorax are inserted above the line.
210. After vitio communi is crossed out.
211. G. Nycteris and Nycterida are crossed out in L. margin and G. νυκτερίς is inserted from R. margin.
212. After 'enim' 'quam vis neque adunas unguibus, neque falcator rostro firmentur' is crossed out.
213. Modern Spanish murciéiago.
214. 'Authores' is inserted above the line.

*Little owl*

The little owl[215] is not equal in size to the ulula, but about the same size as a hen, with a large head. When surrounded by many birds it fights fiercely on its back, sometimes helped by the eagle. It pursues the crow with a special hatred. From the night, it gains the names *noctua* and *nicticorax* (night crow), not unsuitably. For though it has beautiful greenish grey eyes; it is, nevertheless, weak of sight during the day, a failing it also shares with the other nocturnal birds we have recorded.

*Put here the chapter about the bat and the woodpecker.[216]*

Because of this, they lie hidden during the day, and come forth to hunt at night.

*Bat*

This fact brings us to consideration of the bat. For although it is not equipped with hooked talons or a sickle-shaped beak, it can nonetheless in no way withstand the brightness of the sun because of its weak eyes. It flies round at night, in the morning, and in the evening on wings of skin or parchment-like wings. Thus, it is given his name from the evening as was said by Ovid: *They fly at night detesting the light, and get their name from the late evening.[217]*

They have the ears, teeth, teats, head, colour, and hairs of a mouse. They bring forth their chicks live and nourish them with milk, and, as they fly about, embrace their young. From this there was a doubt among authorities examining natural history as to whether the bat should be numbered among the birds or land animals. *Speak here about the woodpecker.*

---

215.   Ken Nicholls suggests that this may be the now extinct great Irish owl.
216.   This note indicates the extent to which this manuscript was a collection of notes comprising a work-in-progress.
217.   Ovid: Meta. Book 4 line 410.

# C. XVIII
## Corvi et monedulae

*I. Corvus*
*G. κόραξ*
*H. Cuervo*
*I. fiaech duibh*

Sunt in Ibernia et aliae aves quamvis rostro minime falcatae, rapto tamen plerumque viventes: ut corvi, qui colore.nigerrimi sunt. Unde Martialis in lentinum canos capillos nigro colore inficientem lepide iocatur.

*Mentiris iuvenem tinctis Lentine capillis*
*Tam subito corvus, qui modo cygnus eras*

*Examinenter hic melius*

In arboribus celsis, atque rupibus nidificant: foeminam ovis incubantem mares pascunt: ad quinos pullos excludunt, teneros etiamnum nidis expulsos volare cogunt, robustos longius fugant: nam ipsi parvis in vicis vix, plusquam singula coniugia degunt: ubi cadavera futura sunt, eo triduo, vel certe biduo ante caedem saepe volare, vulgo traduntur.

Eo minores frugivore eodem colore rostroque directo, sed instar ossis albo durissimoque armatae. Confertissimis[218] agminibus confluentes, ni arceantur, segetibus non nihil detrimenti afferunt, grana comedendo.[219] In Corvorum genere reponunt scriptores non modo frugivoras, sed et aves, quae dicentur monedulae a monetis furandis, quod nummos, fibulas, et id genus avidissime surripiant. Earum in Ibernia species novi,[220] Gracculum, Picam.

*L. Frugivora*
*I. Cnaibhfhaich*

*Hic est graculus cnaibh fhiach H. grajo de quo infra Mondedula locentur infra*

Cornix frugivora minor, columba maior, colore dorsum caerulea, pennas summas, caudam extremam, caput atra[221] voce garrula, carne et alio pabulo vescitur. Cum alioquin sagax videatur: quippe quae nucem duritie rostro repugnantem

*L. Cornix*
*H. Corneja*
*I. fionnoig Chorraich*

---

218.   Quae is crossed out before 'confertissimis'.
219.   Asecondem is crossed out in 'comedendo'.
220.   Tres is crossed out after 'ibernia' and 'cornicem' is crossed out after novi.
221.   After atra 'embellies aurutas nuncupatur' is crossed out.

# C. XVIII
## Crows and Jackdaws

*Crow*

There are in Ireland also other birds who live mostly on what they capture although they do not have hooked beaks, such as the crows who are very black in colour. Thus, Martial jokes charmingly against Lentinus, who is dyeing his grey hairs black.

*Mart. 3, 43*

*You pretend to be a young man with dyed hair, Lentines,[222] so suddenly a crow where recently a swan.*

*Let these be examined more carefully.*

They nest in high trees and cliffs; the males feed the female while she incubates the eggs: they hatch about five chicks at a time; they drive their chicks while still young from the nest force them to fly; when they are strong they drive them further away. For they themselves live in small communities, and rarely have more than one mate. It is a common tradition that where animals are about to die, three days, or for certain two days, before their death the crows will often fly there.

Smaller than the crow is the rook, of the same colour and equipped with a straight beak, which is white and very hard like a bone: they come together in very dense flocks and unless they are kept away they do a lot of damage to corn crops by eating the grain. In the family of crows, writers place not only rooks and the birds that are called jackdaws, because of their habit of stealing money – they steal coins, brooches and that type of thing in a most avid manner.[223] I knew three species of them in Ireland: the carrion crow, the jackdaw, and the magpie.

*Carrion crow*

*This is the jackdaw discussed below*

*Jackdaws are placed below*

The carrion crow is smaller than the rook but larger than the dove. It has a dark blue beak, but the upper feathers, the tip of the tail, and the head are black. Its voice is garrulous. It feeds on flesh and other food. Although in other respects it may appear wise, in as much as, because of its hardness, it will break a nut that

*Carrion crow*

---

222. In the original quotation it is 'laetine' not lentine.
223. There is a Latin pun on jackdaw 'monedulae' and money 'moneta'.

*Dic hic de Screachog*  in togulas atque lapides ex alto saepius iaciendo frangat, eo stolida habetur, quod nidum non satis abscondat, in obviis locis collocando: [dic hic de screachog][224] volantes tamen pullos donec robore firmi evadant, pascit.

Coniugii legem mira concordia servat, sine contentione, sine querela.
Cum coniux moritur, superstes in viduitate vitam agit.

*dic hic le monedula si videtur*[225]  Cum cornice corporis magnitudine comparandus gracculus, *L. Graculus* nigro corpore, rostro saepe rutilo, et rotundo, quando cicur *H. Grajo* fit domum redit: garrulitate importunissimus: unde gracculus *I. Cnaibh Fhiach*[226] inter Musas proverbialiter is dicitur qui insipiens documenta sapientibus loquacissime et ineptissime tradit.

*L. Pica*  Huic loquacitate cedens, expressione verberum pronuntiatione
*H. Pega. Hurraca*[227]  praestat
*Ib Caig. Breach*  Pica, sermonis humani imitatrix optima teste Martiale.

*In xeniis*  *Pica loquax certa dominum te voce saluto:*
*Si me non videas esse negabis avem*

Gracculo minor,[228] nigra est albis punctis variegata. Cum nidum ab homine diligentius visum sentit ova transfert alio: quae quoniam digitis complecti non potest utitur artificio miro: namque surculo super ova bina posito alvique gluttino ferruminato,[229] cervicem mediam subdens aequa utrinque libra, ea portat.

---

224.   This insertion is in a different hand.
225.   This is inserted in the L. margin in a different hand.
226.   Caig is crossed out before I. Cnaibh Fhiaich.
227.   Modern Spanish urraca
228.   Magnitudine is crossed out and minor inserted above the line.
229.   'Alvique gluttino ferruminato' is inserted from L. margin.

*Speak here about the skreachoig.*

resists the beak by throwing it on to roof tiles and stones from on high, nevertheless it is held to be foolish because it does not hide its nest well enough, as it puts it in obvious places. It feeds the chicks even when they can fly, until they are strong enough to leave the nest. It observes the law of marriage with wonderful harmony, without disagreement, and without complaint. When the partner dies, the survivor lives its life in widowhood.

*Jackdaw*

The jackdaw is comparable to the crow in its body size, its black body, and its reddish round beak. When it is tamed it returns home; it is most demanding with its chattering: thus, the foolish man who gives lessons to the wise in a most loquacious and inept way, is proverbially called a jackdaw among the Muses. *Speak here about the jackdaw.*

*Magpie*[230]

Though yielding to the jackdaw in its loquaciousness, the magpie excels in the vividness and the pronunciation of words, being the best imitator of the human voice, as Martial testifies:

*Martial, 14. 76. in the Xenia.*[231]

*I, the chattering magpie, salute you my lord with a sure voice.*
*If you did not see me you would deny that I am a bird.*

*See jackdaw. Lycos, lupus H. Arandajo*

The magpie is smaller than the jackdaw. It is black and dotted with white spots. When it feels that its nest has been observed too carefully by a man it transfers the eggs elsewhere; and since it cannot encircle them with its claws it uses a wonderful trick: it places a twig over the two eggs and sticks it on to them with wax from its belly, then carries them by putting its neck in the middle underneath them and balancing equally on both sides.

---

230. Ken Nicholls notes that the magpie was not introduced into Ireland until 1688, nor is the description particularly apt, hence some other bird must be intended.
231. Actually the Xenia are in book 13.

# C. XIX
## Aves maximae

*Phasianus*

Quas retulimus hactenus, aves quia vel apud nos assidue versantur, vel sua ferocitate commendantur, vel praecipua natura[232] tenebris videndi, luce caecutiendi mirandae sunt: ex ignotioribus haudquaquam putantur. Quamobrem ex earum notitia in aliarum cognitionem lectorem[233] brevibus documentis facile ventrum, puto: si maximas quasque huius insulae aves, quantum collocationis ratio permiserit, primas recenseamus.

*L. Grues*
*G. γέρανος[234]*
*H. grulla*
*I. Coirri moini*

Itaque sunt hic frequentes grues grandis aquilae magnitudine, colore dorsi[235] ex viridi cinereoque mixto, pectore albido, summis alis subnigris, longissimo collo, rostro firmo, recto, acuto, et minime brevi, cruribus altis, digitis quatuor.

*Dic de Regina et eius comite in grues versis[236]*

Mansuefactae ita simplices, et stultae, ut hominem praetereuntes, et sequentes gyros faciant: fere usque adeo solertes, ut disciplinae militaris partim[237] inventrices fuisse videantur. Interfacturae confertissimo agmine conveniunt: quo tempore proficiscantur consentiunt: ducem quam sequantur, eligunt: quae erecto collo providet, atque praedicit: in extremo agmine, quae acclament, et gregem voce contineant, collocatas habent: ad prospiciendum alte volant Marone test

*... quales sub nubibus atris*
*Strymonidae dant signa grues atque aethera tranant*
*Cum sonitu, fugiuntque notos clamore secundo.*

---

232.   This is not quite clear in the Ms and may be nata but natura makes more sense.

233.   'Nos' is crossed out and 'lectorem' is inserted above the line.

234.   G. Geranos is deleted and G. γέρανος is inserted next to it in the L. margin.

235.   Dorsi is inserted above the line.

236.   'Dic de Regina' is inserted in L. margin in a different hand.

237.   'Partim' is inserted from the R. margin.

# C. XIX
## The Largest Birds

*The Pheasant*

The birds which we have reported up to this point because either they live among us constantly or are marked out by their ferocity or are to be wondered at because of their special faculty of seeing in the dark and being blind in the light, should in no way be considered among the less well-known birds. For this reason I think that the reader, from his knowledge of these, will easily come to knowledge of other birds by means of short examples, if we review first all the biggest birds of this island, in so far as the system of classification permits.

*Crane*

On this island cranes[238] are frequent. They are the size of a large eagle, the colour of their back is between green and ash coloured, they have a white breast, the tips of the wings are almost black, their neck very long, their beak strong, straight and sharp and very long, and their legs long with four toes.

*Speak here about the queen and her companion turned into cranes.*

When they are tamed they are so simple and stupid that they wheel about going ahead of and following a man: they are almost clever enough to appear to have been in part the discoverers of military discipline. When about to go on a journey they come together in a very dense line. They agree together when to set out, and they select a leader to follow.[239] The leader looks ahead with head held high and calls first; at the end of the line, they have birds assigned to call and to keep the flock together by their cry. They fly high up for the purpose of seeing ahead as Maro testifies:

*Virg. Aen.10. 265–66.*

*. . . just as under the dark clouds the cranes of Strymon*
*give signals and pass through the aether with a noise and they*
*flee the south winds with a propitious cry.*

---

238.   The grey crane.
239.   Pl. 10, 30, 52–59.

Noctu excubias disponunt: quae vigilias agunt, lapillum altero pede sustinent:[240] lapillus somno laxatus decidendo, et indiligentiam coarguit, et excubantem excitat. Caetera pedibus alternis insistentes, capite subter alam condito, dormiunt.

Grue ciconia[241] corporis quidem molle minor,[242] crurum tamen gracillimorum altitudine, cervicisque longitudine par, rostro longiore praestat, colore alarum nigro, caetero albido e[244] celsis arboribus, et nonnumquam turribus[245] vocem edere solet, ut author Philomelae meminit:

*L. Ciconia*
*G. πελαργός*[243]
*H. Ciguena*
*Ib. Coirrigreni Ieski*[246]

'*Glotorat (sic) immenso de turre ciconia rostro*'

Lastrandi causa lychnitem[247] gemmam, quam quidam remissiorem carbunculum esse, dixerunt, in nido collocare fertur. Ingeniosa, prudens icta admoto vulneri origano, sese curare dicitur: pietatis exemplum parentes[248] senescentes, et alimento comparando instabilis pascit.[249]

---

240.  Ms 'sustenent' should be 'sustinent'.
241.  Ciconia inserted above the line and 'non multum dissimilis est Ibernia Ciconia' crossed out in the line.
242.  After 'minor' 'par tamen illi' is crossed out. 'Tamen' is inserted above the line, crossed out, and reinserted after 'crurum'.
243.  G. Pelargos is crossed out in R. margin and G. πελαργός inserted from the L. margin.
244.  'In' is crossed out and 'e' inserted above the line.
245.  After 'turribus' 'indificat aquibus' is crossed out.
246.  Ib. Greini crossed out in R. margin and Greni insered above it.
247.  Before 'lynchnitem' 'lychnt' is crossed out.
248.  After 'parentes' 'et filios instabiles' is crossed out.
249.  'ineptos pascit' is crossed out and 'instabiles pascit' inserted in a different hand.

At night they place sentries who keep watch, holding a little stone with one foot: when the little stone is released, through sleep, it exposes the guard's lack of diligence and wakes it up. The rest sleep standing on one foot with their heads hidden under their wings.

The stork has a slightly smaller body than the crane, but is equal   *Stork* in the length of its very slender legs and the length of its neck: it exceeds the crane in the length of its beak. Its wings are black in colour, but the rest of it is whitish; it is wont to let out a cry from the highest tree and sometimes from towers as the author of Philomela mentions:
*The stork cries from the tower with his huge beak.*[250]

For the purpose of purification, it is said to put in its nest a white marble stone, though some say it is a semi-precious stone. The stork is clever and wise, and is said to cure itself by applying wild marjoram to the wound: an example of dutifulness, it feeds its ageing and incapable parents.

---

250.   Glotorat is postclassical and means 'the sound that a stork makes' in *Anthologia Latina* 737.7 (Ed. Riese 1894, 1870).

*de ardea vide Arist.*
*L.9 c.17 et Plin.*
*L.10c. box 23*
*L. Ardea*
*G. ἐρῴδιος*[252]
*H. Garca*[254]
*I. Coirri Greini (Corr*
*Iasc)*
*I. Georg.*

In Ibernia ardea notissima est, ab arduo, i. alto volatu nomen adepta iuxta illud Maronis.

. . . *atque altam supra volat*[251] *ardea nubem.*

*Coleach dubh kearch*
*dhuibh coileach feadh*
*kearc feadh*[253]

255

256

Eius tres species ab authoribus numerantur, Leucon, Asterias, Pellos. Leucon vel alba est colore cinereo ciconiae magnitudine, in utroque pede tribus digitis, simo collo, in celsis arboribus plerumque nidificat, lacuum et maritimas oras piscandi causa frequentat: vociferando[257] pluvias, tempestatesque praenuntiat: ingenio tristis, cauta, sagax, provida, diligentiae symbolum. Huic similas est asterias, vel stellaris, praeterquam quod nigris astris, vel punctis insignitur, cauda breviore, unguibus praelongis: butio nonnullis, ut Philomelae authori vocatur. *Inque paludiferis butio bubit aquis.*

Utrique parum dissimilis Pellos albo colore in coitu angitur: mas cum vociferatu sanguinem ex oculis perfundere; foemina non minus aegre parere perhibetur.

---

251.    The 'r' of ardea is inserted above the line.
252.    G. Erodios is crossed out in L. margin and G. ἐρῴδιος is inserted from R. margin. I. Corri Ieskigh is crossed out in L. margin and Coirris Iaska is inserted over it in a different hand.
253.    'Coileach dubh kearc duibh coileach feadha (four words crossed out), gallian' – this insertion in the R. margin is in a different hand.
254.    Modern Spanish Garza.
255.    Before and after tetrax there is written in a different hand that is difficult to interpret.
256.    In L. margin Bonnain lliana and below this Ian Fionn are crossed out.
257.    'Pluvias' crossed out before vociferando.

*Concerning the heron*
*see Aristotle Bk 9, c.17*
*The History and Pliny*
*Bk. 10, c. 60 and 23.*
*Heron*

In Ireland the heron is very well known. Its name is taken from his lofty, that is, his high flight, according to the famous line from Maro:

'. . . *and the heron flies above the high cloud*'.[258]

*Capercaillie male and female*

Three species of the heron are recorded by the authors: the egret, bittern, and grey heron.[259] The egret is white or greyish in colour, the same size as the stork, with three toes on each foot, and a flattened neck. It mostly nests in high trees, and frequents the shores of lakes and the sea for the purpose of fishing; by a loud cry it predicts rain and storms; it has a sad disposition and is wary, shrewd, and cautious – a symbol of watchfulness. Similar to it is the bittern or starry heron.[260] Apart from the fact that it is marked with black stars or spots, its tail is shorter and it has very long claws. It is called the *butio* by some, like the author of Philomela:

'And the butio cries in the marshy waters.'[261]

*Giraldus Cambrensis says that wood peacocks have been seen in Ireland, etc.*

Very similar to both is the Pellos, white in colour. It is distressed during intercourse: the cock with a cry is said to pour forth blood from its eyes, while the female is said to lay eggs with no less difficulty.

---

258. Virg. G.1 364. The heron is called 'ardea'; the word for lofty is 'arduus'.

259. See Pliny 10, 79, 164.

260. A pun in Latin: asterias or stellaris.

261. Author of the Song of Philomela, C.762 42, p. 249 in *Anthologia Latina*.

*In Xeniis*
*L. Phasianus,*[262]
*Phasiana*
*G.* φασιανός
*H. Faisan*
*Ib Coilleach cruoigh,*
*kearc cruoigh*

Phasianum apud Phasidem flumen iuxta Colchidem Asiae regionem primum visum, et inde ab Argonautis in Graetiam transportatum fuisse Martialis vult:

*Argoa primum sum transportat carina*
*Ante mihi hotum non, nisi Phasis erat*

Quae res quomodocumque sit, liquet in Ibernia magno numero esse phasianos[263] gallinae villaticae[264] magnitudine pares, est meleagidis magnitudine et haec avis plane pulchra, rubro, viridique colore varia, cauda longa, et speciosa, oculos habet grandes atque nigros, rubris villis palpebras vestientibus.

## C. XX
### Aves mansuetarum nominibus vocatae

Ex illis modo, quae cum mansuetis commune nomen sortiuntur, aliquas adiungamus.

*Columbacei generis*
*considerato melius*
*examinada*

Feras columbas docuimus proprie dici palumbes. Palumbes vero sive Palumbus in Ibernia est[265] maior columba, colore pulcher, alius albus, alius caeruleus, alius viridis, alius cinereus, alius varius: in rupibus *ib* atque sylvis habitat. Hoc author Philomelae sensit: *Plausitat arborea clamans de fronde palumbus* illum ad annum trigesimum vivere, sunt authores. Senescentis argumentum sunt ungues longi, et incommodi: sed qui citra perniciem possunt recidi.

*L. Palumbus*
*G.*
*H.*
*Ib Coluir fiain*

Hyeme mutus, vere vocalis est, bis anno parit, ova plerumque terna,[266] duos pullos educit, a meridie in matutinum foemina, caetero mas ovis incubat. Ex huius genere conspicitur hic alius

*L. Torquatus*
*G.*
*H. Torcaz*
*Ib fearan, ferran einn*
*et caoran in Xeniis*

---

262. All the following are crossed out in the L. margin. Haec avis non est phasian ub sed pavo sylvestris Cambrensis. Phasius non est sed pavo sylvestris Cambrensis. *L. Phasianus, G. Frasianos, His. fhaysan lycos, I. Coileach fea, Kearc fhea.*
263. Crossed out in the R. margin. Pavonibus, Sylvestris, Iberniam scatare, ipse Cambrensis prodit.
264. From 'gallinae villaticae' to 'plana pluchra' inserted from the R. margin.
265. 'Est' is inserted above the line.
266. Trina is crossed out and terna inserted above the line.

*heasant*

*Xeniis*[267] *Mart.*

*3,72 Epigrams*

Martial will have it that the pheasant was first seen around the River Phasis near Colchis, a region of Asia, and from there was brought by the Argonauts into Greece.

> *'I was transported first by Argo's keel; before that Phasis was all I knew.'*

However that may be, it is clear that pheasants exist in great numbers in Ireland, equal in size to the farmyard hen and the size of a guinea hen. This bird is clearly beautiful. Variegated with red and green, with a long and beautiful tail, it has large black eyes with red tufts of hair adorning its eyelids.

## C. XX
### Birds called by the Names of Tame Birds

Let us add some of those which share a common name with tame birds.

Wild doves we have said are correctly called wood pigeons. The wood pigeon (Palumbes, or truly Palumbus) is larger than the dove in Ireland; beautiful in colour, it can be white, blue, green, grey, or variegated: it lives in rocks and woods. The author of *Philomela* realised this:

*Wood pigeon*

> *The wood pigeon crying out claps his wings from the woodland foliage.*[268]

*more measured*

*⋯nsideration of the*

*⋯ve species L.*

There are people who say that it lives to thirty. (Long and troublesome nails are an indication of old age, but they can be cut back without causing harm. Silent in winter, it sings in spring. It breeds twice a year, laying mostly three eggs at a time. It raises two chicks. From midday to early morning the hen sits on the eggs and for the rest of the time it is the male. From this family there is seen here another bird, slightly larger, bluish grey in colour and handsome, the neck surrounded by a white

*Turtle dove*

---

267.   Martial, book 13, 72.
268.   Auct. Carm. Phil 762, 21. Plausitat means 'to clap the wings'. It is postclassical. *Anthologia Latina* C.762, p. 245.

paulo maior colore glaucus, atque speciosus, candido plumarum torque collum cinctus, inde torquatus nomen habens, qui[269] multum glandis comedit. Eius carna quae dura est, venerem cohiberi, tradidit Martialis.

*Inguina torquati tardant, hebetantque palumbi*
*Non edat hanc volucrem, qui cupit esse salax*

*L. Turtur*
*G. τρνγών*
*H. Tortola[270]*
*Ib turtuil*

In genere columbani minimi censentur turtures columbae moribus similes, insignes fide in coniuges, quibus e medio sublatis, sese perpetuae viduitati, luctuique addicere, nec ad alias nuptias transeuntes nec in viridi ramo subsidentes memorantur. Quamobrem ab authore Philomelae casti iure nuncupantur.

*Et castus turtur atque columba gemunt*
(*Feris quoque gallinis insula abundant et aquatisis de quibus suo loco dicetur quam agrestibus huc spectantibus*).[271]

Harum otis,[272] sive tarda gallina villatica maior, adeoque gravis, ponderosa, volatu lenta, ut si eam venator in aperto studiose sequatur, consequatur[273] etiam, colore fusca in gratio cibatu ponitur.

*Tarda fertur esse in Iberna*
*L. Tarda*
*G. ὠτίς[274]*
*H. Abutarda*
*I.*

Tarda magis fuscus Cynchramus, qui vetula dicitur, perdicis magnitudine coturnices comitatur, in Ibernia per aestatis finem, et autumni principium apparet, rauco cantu obstrepens, quem homine propius accedente intermittit, nidum in messe, urtica, vel sepibus collocat.[275]

*L. Vetula*
*G. Cynchramus*
*H.*
*Ib treine*

---

269.  After 'qui' 'glandes' is deleted.
270.  Modern Spanish tórtola.
271.  This whole paragraph is inserted from the R. margin and crossed out.
272.  After 'otis' two illegible words are crossed out.
273.  Etiam crossed out before consequatur.
274.  G. Otis is crossed out and ὠτίς is placed after it. I. Kearc fhian? is crossed out.
275.  Crossed out after collocat: 'Gallina villatica minor scolopax qui rusticula longo rostra in Ibernia hyeme primoque vere conspilitur, relique tempore non cernitur. Ibi vere eius nidum vel aven nullus umquam vidit quamobrem ubi nidificat minime constat. Mane, vesperieque prodiens submeridiemque quiescet'.

collar of feathers, from which it has the name torquatus.[276] It consumes a lot of acorns. Its flesh is tough and Martial tells us that it inhibits sexual desire:

*Martial 13, 67.*

*Wood pigeons make the private parts sluggish and weak.*
*Let the man not eat this bird who desires to be lecherous.*

*Turtle dove*

Turtle doves are considered to be the smallest species of dove. Similar in habits to the dove, they are famous for their fidelity to their partners. If the partner dies they remain always single and give themselves over to lamentation. No one mentions their entering into another marriage nor their sitting on a green branch. For this reason they are called chaste and rightly so by the author of *Philomela*.

> *And the chaste turtle dove and the pigeon coo.*[277]

*Bustard*

Of these the bustard, or the shaggy bustard is larger than the farmyard hen, so heavy, ponderous, and slow in flight that if a hunter follows it diligently in the open he also catches it. It is dark in colour and makes pleasant eating.

*The bustard is said to live in Ireland.*

*A description of the corncrake. Check its name. Do not delete tortigometra.*[278]

*Corncrake*

The corncrake, darker than the bustard, is called a vetula (old woman). It is the size of a partridge and accompanies the quail. In Ireland it appears at the end of the summer and the beginning of autumn, making a clamour with a raucous song which, when a man comes close by, it discontinues. It places its nest in standing corn, stinging nettles or hedges.

---

276.  Meaning 'with a collar or necklace'.
277.  Auct. Carm. Phil C.762. 20 *Anthologia Latina*, p. 248.
278.  O'Sullivan Beare had deleted this section in the original manuscript. The above marginal note is in a different hand, adding the word 'tortigometra' to provide the Spanish name that the author had omitted.

*L. Rusticula*[279]    Scolopax aut rusticula; perdici similis praeterquam quod est
*G. σκολώπαζ*    longo rostro, colore perdici similis, eius caro in cibum probatur.
*H. Histana Perdiz*    Rustica perdix nonnullis vocatur iuxta illud Martialis:
*pardella gallinella*    *Rustica sum perdix: quid refert, si sapor idem es*
*I. Creabhuir*    *Charior est perdix, si sapit illa magis*[280]
*(Creabhar)*

*Dic hic de phasiand*    *Creabhuir inserted in a different hand. In Xeniis dic hic de Phas.*
*anos.*

# C. XXI
## Aves minores

*H. Abubilla*    Iberniae upupa colore varia, lineis nigris, fuscis, albis distincta,    *L. Upupa*
*I. Fuilkeog*    plumeum apicem in vertice habens, homine ad nidum    *G. ἔποψ*[281]
*L. Perdix*    appropinquante prae timore vociferatur, ad cuius voces aliae    *H. Abubilla*
*G. πέρδιξ*[282]    frequentes, ut in auxilium fluunt; et alium atque alium locum    *I. Fuilkeog*
*H. Perdiz*    occupando aucupem deceptum a nido conatur avertere.
*I. Patrice*

---

279.    G. Scolopax crossed out, H. Gallina cica? crossed out. I. Creamuir?
and Ana Neaska crossed out, other words illegible in Xeniis.

280.    Crossed out after magis: 'Hic gallinarum genere species adscribuntur
pavonibus autem sylvestrius Iberniam scatare Gyraldus Gambrensis
prodit. Sylvestris esse pavo gracis aex, idest capella et hanesto? Dicitur
atque columbae magnitudine et avis garrula et quam pulcherrima,
capite in quo plumeus apex surgit, ventre, humeris albis, collo viridi
et folgente dorso quoque viridi rubris punctis notato, alis partim
viridibusque partim caeruleis praeterquam quod longissimarum
pennarum exterminates, nigerriato colore inficiuntur quamubrem
colorum viarietatem nonnullis varia vocatur. Praeterea rostro longo
nigroque et cruribus altis venientibus nidis extendere dicitur. Haec
gallinarum'.

281.    G. Epops not crossed out in R. margin but 'ἔπαφον and ἔποψ' inserted
in L. margin. I. Phillibbin miog? crossed out above Fuilkeog and an
illegible word, crossed out after Fuilkeog in the right margin.

282.    G. Perdix crossed out in L. margin and G. πέρδιξ inserted from the R.
margin.

*Woodcock*

*Speak here about the pheasant.*

*Concerning the heathcock which in Spanish is called the Ganga.*

The woodcock or rusticula is similar to the partridge apart from the fact that he has a long beak, and is similar in colour to the partridge. Its flesh *or* is valued as a food. The woodcock is called a partridge by some, according to Martial: *I am a woodcock. What does it matter, if the taste is the same? The partridge is dearer, thus it tastes better.*[283]

## C. XXI
### Smaller Birds

The Irish hoopoe is of variegated colour, decorated with black, *Hoopoe* dark, and white lines, and with a crested peak on its head. When a man approaches the nest, it cries aloud through fear. To its cries others in numbers fly together as if to help; and it tries to divert the deceived fowler from its nest by occupying now one place now another.

---

283.  Martial, 13. 76.

*L. Perdix*
*G. πέρδιξ*
*H. Perdiz*
*I. Patrrisc*

Perdices in plurimis regionibus inveniri, sed ut sunt regiones diversae, sic fere perdices distingui, scriptores memoriae prodiderunt. Iberniae perdix columba minor, merulo maior, colore cinereo,[284] punctis misto,[285] cyaneis cruribus, brevique rostro in deliciis habetur olim quoque quanto perdix fuerit in pretio, testatur // Martialis.

*Ponitur ausoniis avis haec rarissima mensis*
*Hanc in lautorum mandere saepe soles*

Eadem fere magnitudine gignitur in Ibernia attagem, sive attagena, colore prope cinereo, nigris punctis variegato: maiore macula nigra pectus hyeme inficitur, quam aestate non habet. Est pinguis, atque sapore gratissimus, annuente Martiale.

*Inter sapores fertur alitum primus*
*Ionicarum gustus attagenarum*

*L. Attagen*
*G. ἀτταγᾶς*[286]
*H. francolin*
*I. fiodog*

*L. Galgulus*
*G. ἴκτερος*
*H. Galgulo*
*De Galgulo Plinius*
*c. 25*

Attagini[287] magnitudine par, plus minus, galgulus, artificio miro pedibus ex surculo suspensus capit somnum, videnti contra morbum ictericum opitulari fertur.

(Inserted from R. margin) Et quae cum foetum exudere abeunt galgulae et upupae eic c. 33. Galgulos ipsos dependents pedibus somnum capere confirmatur quia tutioris ita se sperent, 1. 30 c.11. (Inserted from L. margin) De Galgulo Plin.1. 30 c.11 avis icteris vocatur a colore quae si spectatur sanari id malum tradunt et avem mori hanc puto latine vocari galgulum.

Picus Martius in Ibernia frequentissimus est, avis variis coloribus, albo, nigro, rubro, caeruleo pulchra, cauda longula,

*L. Picus Martius*[288]
*vel Arborius*
*G. δρυοκολάπτης*
*H. Pico*
*I. Snaic brech ac buic*

---

284. After cinereo two illegible words are crossed out.
285. Misto is inserted above the line.
286. Both L. Attgen and G. ἀτταγᾶς are crossed out in R. margin but not replaced in L. margin.
287. Attagine minor sed merulo maoir is crossed out before picus martius.
288. L. Picus Martius and G. Relecas? are crossed out and the new L. and G. names placed above in margin. I. Breac is crossed out before breac.

*artridge*
*Martial 13, 65*

*the Xenia.*
*Martial 13, 65*

Writers have recorded that partridges are found in many regions but that just as the regions are diverse, so are partridges. The Irish partridge is smaller than a dove, larger than a blackbird, ash coloured, spotted, with dark blue legs and a short beak: of old it was also considered a delicacy. Martial testifies how high it was in price. 'The price is rarely served on Italian tables often you will place it on the tables of the sophisticated.'
Martial testifies how high it was in price:
*This bird is rarely served on Italian tables. Often you will eat it at the tables of the wealthy.*[289]

The heathcock, almost of the same size, is native to Ireland. It is almost ashen in colour, it is called attagen or attagena, dotted with black spots: the breast is tinged with a larger black spot in winter which it does not have in summer. It is fat and with a most pleasing taste as attested to by Martial:  *Heathcock*

*Martial 13, 61.*

*Among the flavours of birds the most outstanding is said to be the taste of Ionic heathcock.*

*olden oriole*
*oncerning the*
*olden oriole, see Pl.*
*0, 25.*[290]

Equal in size, more or less, to the plover is the golden oriole; by a wonderful cleverness it sleeps suspended by its feet from a twig; it is held that it protects from jaundice those who spot it.

*When they have raised their young they go away. The oriole and the hoopoe etc. c.33. It is indeed established that the golden oriole takes its sleep hanging by its very legs because it hopes thus to be safer. Concerning the golden oriole Pl. 30, 41. There is a bird called jaundice (icteris) from its colour. If one with jaundice looks at it, he is cured from that disease and the bird dies. I think the bird is called* galgulus *in Latin.*[291]

*oodpecker*

The woodpecker is very common in Ireland, a bird of varying colours, white, black, red, blue; it is beautiful, with a long tail, and a strong beak and hooked nails with which it hollows out trees, and in their hollows it builds its nest. Pliny refers to a

---

289.  The usual quotation for the second line is *hanc in piscina ludere saepe soles*. 'You will often see it playing in a fish pond.'
290.  This reference is inaccurate; the correct reference is Pliny, 30.94.
291.  O'Sullivan Beare gets the name *galgulus* wrong. *Galgulus* is a witwall but O'Sullivan here mistakenly considers it is a golden oriole or ἴκτερος.

firmis rostro, et uncis[292] unguibus quibus arbores cavat, in ipsarum cavis nidificat. Cavis adactos cuneos admota a pico quadam herba elabi,[293] vulgo credi refert Plinius herbam posse inveniri, si sub nido cuneo adacto, culcitra sternatur, aliqui opinantur;[294] quin non modo ceneum; sed etiam clavum quantalibet vi infixum arbori, in qua nidum habeat statim exilire, cum cuneo, vel clavo Picas insederit, Trebius, apud Plinium author est. L. 10, c. 18.

Aliam pici speciem numerant scriptores luteo colore. Lingua acutissima, et firmissima Torquillam nominantes, quod collamquaquaversum facile torqueat. Aristoteles addit,[296] ab hoc arboresfortiter tundi ut culices, atque vermes exeuntes capiat, et eius tria genera facit, alius non multo gallina minus, aliud maius merula, aliud eadem minus.

*L. Torquilla*
*G. ἴυγξ*[295]
*H. Tuercecuello*

*L. Merops*
*G. μέροψ*
*H. Avejurogo*

Equidem nondum habeo compertum, merops, quae et apiastra dicitur, sitne in Ibernia? (Cum tamen eius pabulum[297]), apes,[298] ibi abundare non ignorem.[299] Nihilominus ex Plinio, (eam hic pingere putavi),[300] esse pallido intus colore pennarum, superne cyaneo, inferne subrutilo; in specu sex pedum altitudine defossa nidificare: genitores suos reconditos mira pietate pascere:

---

292. Uncis is inserted above the line.
293. After 'elabi, herbam' is crossed out and 'vulgo cred, referet plinius, herbam' is inserted above the line.
294. 'Vulgo creditur' is crossed out and 'aliqui, opinantur' inserted above the line.
295. 'G. Loligga?' crossed out and 'G. ἴυγξ' placed above it. His.? Turcevello crossed out above H. Tuercecuello. There is no Irish name after.
296. 'Aristotles adit' to 'radem mino' is inserted from the L. margin.
297. 'Cum atmen eius pabulum' is crossed out in text but is needed to understand the sentence.
298. Crossed out after 'pabulum' 'cum ibi esse in avium cognitone haud parum versatu mihi negavit, cum tum eius pabulum apes'.
299. 'Eam' is crossed out before 'ex plinio'.
300. 'Eam hic pingere' is inserted from the R. margin.

common belief that wedges that are driven into their holes are made to slip out, with a certain type of grass applied by the woodpecker. Some maintain that if a pillow is spread under the nest, into which the wedge is driven, it is possible to procure this grass. Trebius, in Pliny (10. 20. 40), claims that not only a wedge but even a nail which has been fixed, with as much force as you like, into the tree in which it has his nest, straight away they spring out when the woodpecker sits on the wedge or nail.

Writers record another species of woodpecker, saffron yellow *Yellow woodpecker* in colour, with a very sharp and very strong tongue. They call them a Torquilla (wry neck) because they can turn their necks easily inwhatever way they like. Aristotle adds the fact that it strikes the trees strongly so as to capture gnats and worms as they come out, and he distinguishes three types, one not much smaller than the hen, another larger than the blackbird, and another smaller than the blackbird.

*ee eater*
*1.1.10 c.33 correct*
*f P1.1.50.99*

I have not yet found whether the bee-eater, which is also called the apiastra, exists in Ireland. However, I am not ignorant of the fact that bees (which are its food) are abundant there. Nevertheless, I have decided to describe it here in Pliny's account. The inside of its wings is a white colour, but they are dark blue on top, and underneath reddish. It rests in a cave hollowed out to a height of six feet, hides its parents and takes care of them with wonderful devotion. To these facts one should add from other writers that it has a long hard pointed beak and is almost equal in size to the blackbird.

quibus ex aliis addendum est, esse rostro longo duro, dentato, et merulo magnitudine fere[301] parem.

*PL. 10.42.80*
*L. Merulus*
*G. κόττυφος*[303]
*H. Merla*
*Ib Loin*

Singulis (si bene memini)[302] avibus, quas hactenus ex professo descripsimus minorem quidem generat Ibernia Merulum satis tamen notum tum rostro pallido, tum colore corporis nigerrimo, tum maxime suavitate cantus. Unde per apposite cecinit author Philomelae:

*Et merulus modulans tam pulchris concinit odis*
*Nocte ruent tamen carmina nulla canit*

Non nunquam albus etiam meruluus visus est in Ibernia, frequentiorque solitarius alia meruli species corpore minor.

*Dic de lietraisc et*
*smolach.*
*L. Turdus*
*G. κίχλη*[304]
*H. Tordo*
*I. Lietraisc vide nebriss*
*Arist. Plin.*
*L. Perla*[307]
*G. ἰλλας Vide tordo*
*[L. Martinus pescator*
*G. Ispida*
*H. Martin Vescator*
*Ave del parayso*
*I. Kearc Iski*[308]*]*
*All in brackets deleted*
*in the left margin*

Merulo Turdus Iberniae magnitudine par, nigris et albis punctis variegatus, rubris etiam maculis alas insignis, loquacissimus ex nido vociferatur. Haec avis Martiali gratissimum opsonium iudicatur

*Inter aves turdus, si quis, me iudice certet*
*Inter quadrupedes gloria prima lepus*[305]

Ac in Ibernia[306] quidem turdi sese satis celebrarunt anno Christi Redemptor millesimo, sexcentesimo vigesimo primo cruentis pugnis inter se commissis cadendo, sicuti fusius alibi demonstrabo.

---

301. Fere is inserted above the line.
302. 'Si bene mimini' is crossed out above the line and reinserted in brackets.
303. Kopischos deleted.
304. G. Kichlae is deleted and κίχλη placed above it. I. Smolach crossed out and grues (cranes) inserted above it. Lietraisc is inserted at the side.
305. Crossed out in Ms 'minorem Paulo pelaram vel dortram genus an etiam turdi faciunt scriptores eius meminit author philomela'. 'Dulce pelara sonat quam dicunt nomine dortram sed fugiente die nempe quieta silet.'
306. 'Ac in Ibernia to demonstrabo' is encircled with a line after sturnus but as it refers to turdus I have put it here. There is an insertion mark 'colloca suo loco'.
307. I cannot decipher the Spanish and Irish. Pearla is an unknown bird.
308. Or iski crossed out. 'Hiberniae hispadam ie matinam piscatorem iste Gyraldus martinetam voct, et inter aves quas paucissim as fusius describit, esse merulo maiorem, corpore in coturnicis? Brevi, albo ventre, nigro dorsu, cauda brevi et nigra, ad exigus pisculos capiendos sese in mari mergere mortuam si in sico loco servetur, minime putrescere, si inter vestes siccas collotetur eas et odorare et a tinea defendere, si in sicco a rostro suspendatur, renascentibus plumis (quod ego sospectum habeo), reviviscere'.

*Blackbird*
*Pl. 10,42,80*

Ireland begets the blackbird, smaller than the individual birds which up to this point I have, if I remember rightly, explicitly described, but well enough known both for its pale beak and for the very black colour of its body, but mostly for the sweetness of its song, about which the author of *Philomela* wrote most: *And the blackbird piping sings such beautiful songs. At nightfall however he doesn't sing any song.*[309]

Sometimes the white blackbird has also been seen in Ireland, more frequently alone. It has a smaller body than the other species of blackbird.

*Thrush*
*peak about the field-*
*are and the thrush.*
*Nebrissa*
*Pliny*
*Aristotle*

The thrush, in Ireland, is equal in size to the blackbird, with black and white spots, notable also for the red spots on the wings, it is most vocal and calls out from the nest. This bird is judged to be a very tasty morsel by Martial:[310]
*If anyone should dispute it, in my judgement among birds the thrush is the first glory and among quadrupeds, the hare.*

And in Ireland the thrushes made themselves pretty famous in 1621 by dying after starting bloody fights among themselves, as I will demonstrate at greater length elsewhere.

---

309. Author of the song of Philomela,. C.762, 13–14 in *Anthologia Latina*, p. 247.
310. Mart. 13,92,1.

Merulo corporis molle par, turdo colore similis sturnus loquax       *L. Sturnus*
et hominis et aliarum avium voces imitatur, conferto agmine       *G. ψαρός*[311]
incedit.[312]                                                             *H. Esturnido*
                                                                         *I. Druid breac*

Merulo quoque par coturnix est in Ibernia, colore fusco, punctis
nigris distinguentibus, albido pectore.

*L. Coturnix et qualea*
*ὄρτυ inserted*
*H. Cordoniz*
*I. Gearra guirt*
*L. Cassita Galerita,*
*alauda*
*G. κορυδαλός*
*Cory dos crossed*
*out I. Fuiseog Vide*
*annotationes*

*Dic de chalandra,*
*alauda & smolach*

Coturnici similis, ea tamen minor cassita, sive galerita in Ibernia       *L. Chalandra*
passim conspicitur cono, vel crista plumea quae a fronte in coni       *G. χαλαηδρα*
modum surgit, insignis. (Ei similis est calandra, sed sine apice       *Kalandan &*
suavi vocis melodia commendatur).[313] Volans in sublime, veluti       *Calandrio*
telum sursum iactum, plurimum surgit, et iterum quasi lapis       *I. Faithloig*[314]
gravitate premente cadens, contractis alis descendit, ita tamen
librato corporis pondere, ut se pedibus excipiat.

Hac minor hirundo carne vescens, hybernis mensibus abit in       *L. Hirundo*
Apricos montium recessus, ubi deplumis, atque nuda traditur       *G. χαλανδρα*
inveniri, redeunte vere revertitur.[315]                                 *H. Golondrina*
                                                                         *I. Faithloig*

---

311.   G. Psaros crossed out G. ψαρός inserted, Ib Druid breac crossed out
       here. L. Galgalas, G. Icteros, H. Galgalo, I.
312.   Crossed out in text 'attagini magnitudine par lus minus, colore luteus
       galgulus, artificio mirou pedibus ex surculo suspensus capit somnum,
       videnti contra. Morbum comitialem opitulari fertur'. I could not find
       a reference to this.
313.   The sentence in brackets is crossed out in the text.
314.   These are crossed out in the R. margin.
315.   Chelidon is crossed out.

*Starling*

The starling is equal to the blackbird in size of body, similar in colour to the thrush. It is loquacious, and imitates the voices of men and of other birds. It proceeds in dense columns.

The quail is also equal in size to the blackbird in Ireland, dark in colour, adorned with black spots, and with a whitish breast. *Quail*

The lark (cassita or galerita) is similar to the quail, but smaller. *Lark* It is seen widely in Ireland; it is remarkable for a cone or a feathered crest which rises up from its forehead in the shape of a cone. Similar to it is the *calandra*,[316] but without the crest. It is marked out by the melody of its sweet voice.[317] Flying into the sky like a javelin that has been thrown upwards, it soars aloft and again like a stone falling by the pressure of its own weight, closing its wings, it descends, but with his body weight so balanced that he settles himself on his legs.

*peak here about the halandra, the alauda, nd smolach.*

*Mention here Ib. ortan (unknown) Drepnis & Ocyn*

The swallow is smaller than the lark, and is a carnivore. It *Swallow* goes away in the winter months to the sunny recesses of the mountains where it loses its feathers, and is said to be found naked. On the return of spring it comes back again.

---

316.  Καλανδρός – lark.
317.  The words in brackets have a faint line through them in the text.

*Avion Ib portan*
*L. Cipcellus*
*G. κύψελλος*[318]
*H. vincepe*
*Appraxaque*

Eius altera species cypsellos, sive apodos, brevioribus pedibus ad authoribus memoratur. Ocyn et drepanim addit Plin. L. 15, c. 47, vide, Ib. Portan. et Hispe Avion.[319]

## C. XXII
### Aves minimae

(Circundine minor)[320] sequitur Parus, (cui de Iberno loquor color ex viridi, fuscoque componitur: in spinis et aliis arboribus plurimum nidificat, unguibus tenax),[321] voce minime suavis, ut author Philomelae cecinit.

*Parus enim, quamvis per noctem tinnitet omnem*
*At sua vox nulli iure[322] placere[323] potest*

*L. Motacilla*[324]
*G.*
*H.*
*Aguzanicue neuereta*
*coneta pezpita*
*I Glasoig*

Iberniae motacillia[325] non est praetereunda a perpetuo fere longulae caudae motu nomen habens capite pectore cauda summis alis nigris, caetero argenteo colore. Unam[326] vidi capite,[327] et pectore pallido. Nebr. chirivia. Motacilla, sisura, sisopygis.

---

318. G. Cypcellos is crossed out and G. κύψελλος inserted above it.
319. Plin. 10, 55, 114.
320. Crossed out in L. margin: L. Vitidis, G. Aegithalos, H. Chamariz Gaffaran, I. buoig lin, Gioboig lin, spideoig lin. Crossed out in R. margin: R. illegible, G. ???, H., I. Bolgan Rheoraunn?, & Deach Geharraigh?
321. The words in brackets have a faint line through them in the text.
322. Iure is inserted above the line.
323. There is an illegible word crossed out before potest.
324. G. Cillyros is crossed out in L. margin and G. σεισοπυψὶς in right margin.
325. Inserted from the R. margin. Moticilla magnitudine par buiog kinn oir, aureo capite colore pulchra canora (vide fusius apd Aristoelis is crossed out). The Irish buioc kinn oir means little golden-headed one. The term refers to the goldfinch.
326. From 'unam to sisopygis' is inserted in a different hand after 'argenteo colore'.
327. Capite is crossed out and capit pectore cauda are inserted above the line.

*Swift*

A second species of this is the swift (cypsellos or apodos);[328] with shorter feet, it is recorded by writers. Pliny adds the ocys and the sand martin. Pl. 15c.47 (Pl. 10, 55, 114 correct).

# C. XXII
## The Smallest Birds

*Tomtit*

*Aguzanicue, neue*

*reta, coneta pezpita,*

*sisapaga Nebrissa in*

*ubo?*[329]

(Smaller than the swallow,) the tomtit follows (I am speaking about the Irish one, whose colour is made up of green and black. It nests mostly in blackthorn bushes and other trees). Its claws are tenacious, but its voice is not very pleasant, as the author of Philomela wrote:

*For although the tomtit titters the whole night long, his own voice is not able to please in any way.*[330]

The wagtail of Ireland must not be passed over.[331] It gets its name from the almost constant motion of his rather long tail; its head, breast, tail and the tips of its wings are black while the rest is silver in colour. I have seen one with the head and breast whitish. Nebris, chirvia, moticilla, sisura, sisopygis.

---

328. Apodos actually means without feet.
329. A marginal comment in another hand notes that Nebrissa should be consulted for the names chirivin motacilla. Sisura, sisopygis.
330. Author of the Song of Philomela, 762, 9 & 10 in *Anthologia Latina,* p. 247.
331. Inserted from the R. margin. 'The wagtail is the size of buiog kinn oir (goldfinch), has a golden-coloured head, it is a beautiful song bird'.

*L. Curruca*

*G. ἐλαίς*[332]

*H. curruca*

*I cuacan, cuachan*

*[ finnin feoir is crossed*

*out] coalog, riach*

*L. Passer*[351]

*G. δειρίς στρουθός*

*τρωγλοδύτης*[335]

*H. Garrion Pardal*

*I. Gealbhuin*

Curruca, quod, et cuculi pullos excludit, et ipsum commitatur, satis est Ibernis nota.

Passer[333] qui corpusculo, pedibus et rostro brevibus dorso cinereo ventre albido, mento nigro), qui salax, qui frequentissimis,[334] Ibernis ignorari minime potest, longiore explicatione non indiget.

*L. Carduelis*

*G. ἀκανθίς*[337]

*H. Gilguero colorin,*

*siete colores, cadernera*

*Ib*[338] *Coinilleoir*

*mhuri, finnin oir,*

*finnin feir & finnin*

*feoir quare*

Carduelis a carduis, quorum semina comedit, Latinis dicta, multisque coloribus picta cantus suavitate commendatur.[336] //Plin. L.10 c.29. *Clorion quoque, quo totus est Luteus, hyeme non visus, circa solstitia procedit.*

Veniam nunc ad sylviam, quae Graecis hyeme erithacos, aestate phoenicurus,[339] et Latinis etiam pectore rubro non immerito rubecula nuncupatur: corpore parvo, misericordia magna Ibernis in honoresummo habetur, quod hominum exanima, desertaque cadavera musco,et herbis officiosa, et pia tergere conatur. Quamobrem ea invalet consuetudo, ut eam Iberni non occident,[340] et quamvis canor sit, non petant[341] in caveam, non capiant, quin etiam captam redimant, asserant in libertatem, atque dimittant.

*L. Sylvia, rubecula,*

*Rubicilla G. ἐρίθακος,*

*φοινικουργὸς*

*φοινικυρος πυρροὔλας*

*H. Camachuelo*

*pajarel, rubicilla*

*I. Spideog bruin*

*muintiri suilleabhain*

*zorzal Ib. sacan vide*

*??? et aliacus apud*

*Plin ? c. 46.*

332.   G. Epilais is crossed out in L. margin above ἐλαίς.

333.   From 'passer' to 'mento nigro' is inserted in a different hand and 'passer qui colore cinereo' is crossed out in the text.

334.   After frequentissimis 'et ig' is crossed out.

335.   In fact, the name for the wren. G. Troglito is crossed out and δειρις, στουθος, τρωγλοδύτης inserted from the R. margin.

336.   Crossed out in text: 'Melodia non corpore superior lutea chloris etiam dicitur quod alas, lumbos pectus viridi caeteroquin luteo colore inficitur'. Inserted in a different hand.

337.   G. 'ἀκανθίς' is inserted in the L. margin above two illegible Greek words written in Roman letters.

338.   Ib – This is written in a different hand. Crossed out minutan? in bene buiogfhran & coinlloir crosach, coinlleoiri oir.

339.   G. Erithacus & Phoenicurus crossed out.

340.   'Eam ibernie non occident' is inserted from L. margin. After 'sit' eam tamen iberni non occident is crossed out in text.

341.   'Petant' is inserted above the line and is crossed out after 'caveam'.

Hedge sparrow

The hedge sparrow is well known in Ireland because it hatches out the chicks of the cuckoo and accompanies it.

House sparrow

The house sparrow which in its little body, short legs and beak, ash-coloured back, whitish belly, black chin, is dirty and very numerous and cannot be ignored by the Irish, is in no need of a longer explanation.

Goldfinch

The linnet (*carduelis*) is so called by the Latins from the thistle (*carduus*) whose seeds it eats. It is tinged with many colours and is commended for the sweetness of the voice.
Plin. L.10, C.29. 'The *Clorion* also which is completely yellow and is not seen in winter, comes out around the solstice.'

Robin
spideoig bruin
earaig mhuintiri
uilleabhain (spideog
roinndearg) Zorzal
o. sacan, vide et
diacus avio apud
lin. ? c.46 Alicus
unknown) in Pl. ?
,6 c4.

Let me come now to the Sylvia (the robin) which is called by the Greeks in winter, Robin Redbreast, (*erithacos*) and in summer, robin (*phoenicuros*), by the Latins is also rightly called rubecula because of its red breast. Small in body, it is held in the highest sympathy among the Irish because, pious and courteous, it tries to cover lifeless bodies and deserted corpses of men with moss and grass. For this reason, custom prevails that the Irish do not kill it and although it may be sweet sounding they do not seek it for a cage, nor do they capture it but even when it has been captured they redeem it, and they give it liberty and set it free.

Inde factum puto, ut illa propius, et audentius in Ibernia quam in aliis regionibus ad homines accedat. Eam OSullevana praecipue gens (fortasse quod ex OSullevanis aliquem occisum, atque relictum pro suo more musco operuerit) insignium suorum emblemmati addidit. In hoc pingitur armatum militis brachium cum apro ferarum Iberniae ferocissimo stricto gladio pugnatis, quo summum discrimen significatur: caelantur etiam tres Iberniae cuae, vel venatici canes fugientem cervum sequentes; quibus ob oculos parta victoria proponitur. Ea verba Ibernicum *Foistineach*, id est, '*intrepide*' circumit, monens ne vel in periculo metu premamur, vel in victoria laetitia efferamur, sed utrobique rem intrepide, animo forti, constanteque geramus. Supra omnia rubecula stans exprimitur, hortans ut in victos, miseros, atque calamitosos simus victores pii. Ut omnia nos hoc carmine perstrinximus.

*Urgeat, aut fugiat victus mucronibus hostis,*
*Fortunam intrepide cautus utramque rege.*
*Erithaci exemplo stratorum membra tegentis*
*In victos hostes esse memento pius.*

L. Regulus
G. ῥόβιλλος[342]
βαδιλίσκος Τροχίλος
Πρέσβυς
H. Trochillo.
I. Dreolain

Ob humanitatem tantam, tamquam inauditam erithacus mea quidem sententia potius esset avium rex hominibus appellandus, quam Trochilus, qui rex atque senator vocatur,[343] cum execranda feritate mortuos rostro pugnat: ex exiguus, atque fuscus neque corporis magnitudine, neque colore praestet, neque longo volatu polleat a pueris cursu superabilis[344] avicula parva, et misera, fruteta, et foramina incolit.

*Vide Plin. L. 10 c. 14*
*pag 211 v 22.6. Aris*
*L.9 Hist C. 11 in fin*

---

342. G. Basileus and trochilos are crossed out in L. margin, and the names in Greek lettering are inserted from the R. margin.
343. 'Regulus atque basileos' is crossed out and 'qui rex atque senator' is inserted above the line.
344. 'Omnium fere avicularun minimus atque miserimia' is crossed out after superabilis and 'avicula parva, et misera, frutera e? foramina incolit' is inserted above the line.

As a result of this, I consider the outcome is that it comes nearer and more boldly to men in Ireland than in other regions. Especially the O'Sullivan sept added the robin to the insignia of their crest (perhaps because when one of the O'Sullivans was slain and abandoned, the robin, according to its habit, covered him with moss).[345]

On the crest is painted the armed arm of a soldier fighting, with drawn sword, a boar, the fiercest of the wild beasts of Ireland, by which is signified the highest challenge: engraved also are three Irish hounds or hunting hounds chasing a fleeing stag, before whose eyes, victory having been achieved, is displayed the Irish word *foistineach,* which means 'calmly'. The word goes around, warning that we should not either be overcome by fear in danger, or be carried away by joy in victory but in both situations that we should behave ourselves calmly and with a brave and constant heart. Above all, the robin is portrayed standing, exhorting that, as victors, we should be kind to the vanquished, the wretched and those who suffer calamity. I have touched upon all these things in this verse.

*Manage both kinds of fortune calmly and carefully, whether the enemy presses on or flees vanquished by swords. Follow the example of the robin covering the limbs of the fallen. Remember to be kind to conquered enemies.*

Wren

The robin because of such humanity, and although unreported however in my opinion deserves to be called king of the birds by men rather than the wren, which is called king and senator, since it pecks the dead with its beak with such cursed savagery. It is both small and dark coloured and does not stand out in the size of its body nor in colour, nor is it strong in flight and can be overcome by boys in a chase. It is a little bird and miserable,

*See Pliny book 10, chapt. 14 pag 211 v 22.6; Aristotle History of Animals book 9, at the end of chapt. 11*

---

345.  The crest described above is very similar to that seen in the upper right hand corner of the painting of Donal Cam in his regalia as knight of Santiago. That contains the armed arm of a soldier. Instead of three hounds there are three lions passant, then the deer fleeing, and then a boar. There is a robin, but the shield is surmounted by a helmet with a rayed coronet ontop, and the motto 'Patientia duris gaudet'. In the original painting now in Maynooth College the painting is so dark that people thought that the boar was a bear. However, the copy of that original painting has been restored and cleaned in University College Cork and the animal is clearly a boar.

Eum Iberni pueri cum alias, tum postridie Christi Redemptoris
natalem diligenter inquirunt, captumque in caput suspensum,
et e virga pensilem agunt, quo enim tempore verum orbis totius
regem natum celebrant, ridiculum regulum minimo patiuntur.
Prolis // tamen generatione commendatur, cum ova tredecim
edat, pullos duodecim excludant.

<br>

*L. Salus. Vide Arist*
*& Plin.*
*G. ἀίγιθος*
*H. I. da trian oenin*

Aegitus avium minimus ditatus, pectore albidus dorso alis
longulaque cauda cinereus in Iber. cernitur.

## C. XXIII
### Aves Aquaticae

*Coloca supra suis*
*laocis (entered in a*
*faintly different hand)*

Hactenus postquam domesticarum avium reputationem
absolvimus, de terrestribus avibus, idest, quae in terra duntaxat,
agunt, sedent, et digitos nando inutiles habent, disputavimus.
Iam de aquaticis, quae palmipides sunt, natant, et in aqua
quoque versantur, ac prius de fluviatil ilus dispiciamus.[346]

*L. Anser agrestis*
*G.*
*H.*
*I. Fiagea est anser*
*martinus*

Sunt quidem in Ibernia feri anseres plurimi, domesticis
persimiles. Sunt et[347] species aliquot, domesticis non dissimiles,
ut ipsa, quae fera[348] dicitur, ut hac minor querquedula ut aliae
variae, quas longum fuisset nominare, sive eas anates, sive
gallinas aquaticas appelles. Hae cum fluvios vel lacus, sive dulces
quas frequentent, fluviaticae dicuntur. Sunt vero in Ibernia et
aves, quae maria nando, versunt, ob id marinae nuncupatae.

*L. Anas Agrestis*
*G.*
*H.*
*I. Fia Lacha*
*L. Querquedula*
*G.*
*H.*
*I. Plas Lacha*
*L. Torquata*
*G.*
*H.*
*I. Garrig sraithi*

---

346.   The whole paragraph from 'hactenus' to 'dispiciamus' is faintly crossed
out with a single line.

347.   After et 'anatum ferrarum' is crossed out.

348.   Before 'fera' 'anas' is crossed out.

and it lives in bushes and holes.[349] Irish boys seek it diligently
at other times and also on the day after the birthday of Christ
*the Redeemer* and, when caught, they suspend it by the head and
carry it hanging from a twig. At the same time as they celebrate
the birth of the true king of the whole world but they do not
tolerate the ridiculous wren.[350] It is, however, commended for
the production of offspring since the wren lays thirteen eggs
and hatches out twelve chicks.

*ong-tailed tit*[351]  The aegithus is called the smallest of the birds. It has a whitish
breast, back, and wings, and a longish ash-coloured tail. It is
seen in Ireland.[352]

# C. XXIII
## Aquatic Birds

Up to now we have completed the review of domestic fowl; we   *Wild goose*
have discussed land birds which, as far as applies, lead their lives
and roost on land and have toes useless for swimming. Now let
us look at aquatic birds who have webbed feet and swim and
spend their time in water, first dealing with the river birds.

There are in Ireland very many wild geese very similar to the   *Wild duck*
domestic ones. There are also some species not unlike the
domestic ones, such as that which is called the wild duck, or the   *Teal*
teal, which is smaller than it, and like other different ones that
it would take a long time to list, whether you might call them
ducks or water hens. These, since they frequent rivers or lakes
or fresh waters are called river fowl. There truly are in Ireland
also birds who swim in the sea and spend their time there. For
this reason they are called *sea birds.*

---

349.   Inserted from the R. margin: See Plin. 10,14, also 211, V. 22,6; Arist.
      9 Hist. c.11 in fin.
350.   Inserted from the R. margin: The Latin name for the wren, 'regulus',
      means 'little king'.
351.   Liddell and Scott say small unknown bird, may be linnet.
352.   Ken Nicholls notes that this description seems to fit that of a long-
      tailed tit.

In harum genere mergus non infimum sibi locum vindicat, inde *L. Mergus*
nomen adeptus, quod in aqua sese quam saepissime mergat. *G. Kalikatzo*
Unde cecinit Ovidius: *Aequor amat, nomenque manet, quia* *H. Cuervo marino*
*mergitur, illi*                                                          *Ib Fiach mairi*

Colore, magnitudineque corvo similis est. Aurarum signa
colligens, tempestatemque futuram praevidens, ex aequore
medio tuta littorum praesidia petere fertur. In petris, vel
arboribus nidum condit, incipiente vere parit: pullos ternos
educit, adeo firmos, ut exclusi, si nutricem forte desiderent, ipsi
victum sibi comparent.

*L. Fulicia fulix*[353]    (Mergi species dicitur esse)[354] fulica, sive, fulix, cuius meminit
*G. κέπφος*          Virg. '*Cumque marinae in sicco ludunt fulicae*'
*H. cerceta*
*I. Fuillin chapuill*
*I. Georg.*           Ad idem genus Columbius qui[355] etiam urinatrix vocatur,
                     spectat, tempestate fera, adversoque vento natat ne comedendi
                     quidem causa cursum intermittens, et ovis piscium alitur.

*L. Gavia*           Longe diverso colore, quam mergus est gavia, Graecis dicta
*G. λάρος*[356]      Laros, albo, vel cinereo, corvo minor: in petris nidificat, aestate
*H. Gaviota*         parit ova, ut plurimum terna. Exclusi pulli mare statim petunt.
*Ib Fuilinn*

*L. Alcedo*[357]      Graecis Halcyones, Latinis alcedenes, quasi algedines dicuntur,
*G. ἀλκυών*          quod algido, vel frigido tempore foetus edant. Ante brumam
                     enim diebus septem ad undae oram nidos faciunt, totidemque
                     sequentibus pariunt ova quina. Quo dierum spacio mare
                     placidum, tranquillumque esse solet. Quare dies illi Halcyonides,
                     vel halcyonia nominantur, cuius rei meminit Naso.

---

353.  Cephos crossed out, an illegible word in Greek lettering also crossed
      out.
354.  The words in brackets are crossed out in the text.
355.  'Columbius' is lightly crossed out and 'qui' inserted above the line after
      it.
356.  G. Laros is crossed out. Ib Luch laith is also crossed out.
357.  L. Halcyon is crossed out.

Of this type, the cormorant may claim not the lowest place. It *Cormorant* acquired its name because it very often dives and surfaces in water.[358]

About this, Ovid says: '*It loves water; the name sticks because he dives in it*'.[359]

In colour and in size it is similar to the crow. When they collect together it is a sign of winds. It is held that, foreseeing a future storm, it seeks protection from the middle of the sea in the safety of shores. They build their nest on rocks or trees, they lay their eggs at the beginning of spring, they hatch out three chicks at a time, so strong that when pushed out of the nest, if they should desire nourishment they may provide food for themselves.

*Coot*  A species of the cormorant is said to be the coot (fulica or fulix) whom Vergil mentions: '*Whenever the sea coots play on the dry ground*'.[360]

In the same group belongs Columbius, who is also called the diver, it faces into and swims against the wind in a fierce storm; it interrupts its journey not even for the sake of eating and is nourished by the eggs of fish.

*Gull*  The gull is very different in colour from the cormorant. It is called Laros by the Greeks, and is white or ashen in colour, smaller than the crow: it rests on rocks and lays eggs in the summertime, at the most three at a time. The chicks when they are hatched out seek the sea straight away.

*Kingfisher*  The Greeks call the kingfisher halcyones[361] and the Latins alcedo as if they are being called algedines (cold ones) because they produce their young in the cold or frigid time of year. Seven days before midwinter day they make their nests at the edge of the waves and seven days after solstice they lay five eggs. During that period of days the sea tends to be calm and tranquil

---

358. The Latin name is 'mergus'; the pun here is with 'submerge' and 'emerge'.
359. Ovid. *M.* 11, 795.
360. Verg. *Georg.* 1,363.
361. Alcedo, a(lkuw/n: Alcyone, daughter of Aolus, threw herself in the sea for the love of her shipwrecked husband Ceyx and she was changed into a kingfisher. Ov. *M.* ii, 384.

*Perque dies*[362] *placidos hyberno tempore Septem*
*Incubat Alcyone pendentibus aequore nidis*
*Tum via tuta maris: ventos custodit, et arcet*
*Aolus ingressu, praestatque nepotibus aequor*

Haec avis passere paulo maior, color[363] maiore parte cyanea, purpureis, candidisque pennis admistis, gracilis, proceraque collo,[364] piscibus vivit, subit in amnes, cuius (duo sunt genera Arist. et Plin., minus, et canorum, et fluviatile in arundinetis canit, maius marinum Arist. L. 9 c. 14.)[365] Has Iberniae aves, plerisque scriptoribus, qui hoc argumentum tractant, notas perstrinximus.

## C. XXIV
### Aves Ibernis nominibus dictae

Praeter eas illa insula fert alias multas, quas plerasque alibi quoque reperiri[366] puto; sed nomine signatas non invenio. Aliquas lubet memorare.

Relachae est gallinae villaticae magnitudo, venter albus; pectus et caput, nigrum, rostrum longum acutum, nigrum:[367] reliquus color ex lineis albis, atque nigris textus. Haec anguillis, et aliis piscibus vescitur, in rupibus nidificat, quinque ad summum ova parit, tres, vel quatuor excludit: ipsa, et eius ova esu non ingrata.

Realach (*crossed out 3–4 illegible words Ian fuinn, Alba, avis a colore nominata alba utramque pennam nigra macul. utramqide pennam ificiente, anatis magnitudine minores aviculas insectat?*)

Curlionus eadem fere magnitudine colore gallinae villaticae similis, altis cruribus,[368] longo[369] rostro unius palmi, agros, et littora frequentat, longe volat, pluviam crebro questu praenuntiat.

Cuerlúin Numiniu: *vide?* crossed out
Crotach Mhuire
(*Gubudan deleted*)
Gobodan, Anglice
Godvuinge

---

362. After 'dies' septem is crossed out.
363. 'Passere paulo amplior colore' is crossed out and 'passere paulo maior color' inserted from margin in a different hand.
364. After 'collo' in Ibernia freguons 'est' is crossed out.
365. The writing in brackets is inserted from the R. margin.
366. 'Inveniri' is crossed out and replaced above the line by 'reperiri'.
367. After 'nigrum' 'eudem colore pedes' is crossed out.
368. 'Pedibus' crossed out and 'cruribus' inserted above the line.
369. 'Longissimo' is deleted and 'longo' inserted above the line.

because those days are called kingfisher days or halcyonia, which is also mentioned by Naso.

*For seven placid days in the winter time the kingfisher hatches with its nest hanging in the water. Then the sea roads are safe. Aeolus checks the winds and keeps them from emerging; the sea places a guard over its offspring.*

This bird, little bigger than a sparrow, is bluish in colour for the greater part, and with a whitish and purple mixture on the wings; graceful and with a long neck it lives on fish, it goes up into rivers. There are two types of kingfisher according to Aristotle and Pliny. The smaller and the singing type, associated with the river, sings in the reeds, the greater a sea bird [Arist. L.9.c.14.]. We have touched upon these Irish birds, known to most writers who deal with this matter.

## C. XXIV
### Birds Called by Irish Names

Besides those birds this island carries many others. The most of which I think are also found elsewhere but I do not find them signified by name. I am glad to record some of them.

The oystercatcher is the size of a farmyard hen. It has a white *Oystercatcher* belly, the chest and head is black, the beak is long, sharp and black; the rest of its colours is made up of white and black lines. This bird feeds on eels and other fish; it nests on cliffs and lays at the most five eggs; three or four hatch out. The bird itself and its eggs are not unpleasant to eat.

The curlew is almost the same size and similar in colour to *Curlew* the farmyard hen, with long legs, a beak of one palm long; it frequents fields and beaches. It flies a long way, it foretells rain by a frequent, plaintive cry.

(Huic colore, figuraque similis gobundanus, eo tamen minor eadem loca incolit[370]).[371]

Utroque minor cosderganus .i.[372] Rubripes rostroque breviore; sed sublongo, cruribus rubris, (unde nomen illi faciunt Iberni) altisque, colore gilvo fluminum ripas plurimum visit.

*Coisdearagain an Porphyio?*[373]

*Neska*

(Neska sturmi magnitudine, cruribus, rostroque longulis, colore vario ex lineis nigris, et gilvis in uliginosis commoratur).[374]

# C. XXV
## Aves alias regionibus minus communes

Alias aves, quas ego neque ab authoribus Latinis memoratas observavi,[375] neque alibi vidi, Ibernia nutrit: quarum aliquae nominandae sunt.

Quaedem meleagridis magnitudine, colore subnigro, carnosae, esu non iniucundae, densam sylvam frequentantes usque adeo hominem stupent, ut in arbore sidentes octo, vel novem lapides iactus a viatore spectent, ac saepe non fugiant, donec occisae prosternantur.

---

370. 'Habitat' crossed out and 'incolit' inserted.
371. The words in brackets are crossed out in the text.
372. Inserted from the L. margin: Coisdeargan exalbidi et viridi colore iuxta riuos agit maxime glacie regente. Merulo maior multo minor pica, alischinereus & agit etiam litoribus).
373. Inserted from the R. Margin: Coisdearganus coisdeargain an porphyrio?
374. The text in brackets is crossed out and has a line drawn round it.
375. 'Comperio' is crossed out and 'observari' inserted above the line.

The Crotach Mhaire is a large shore bird of the sandpiper family, *Crotach Mhaire*
it has an upturned bill. The Gobadan (godwit) similar to it in *Gobdan (gobadán, in*
colour and shape, but smaller than it. It inhabits the same places. *English godwinge?)*

The red shank is between whitish and green in colour. It goes *Is the redshank the*
near rivers, mostly in cold weather. It is much bigger than *porphyry?*[376]
the blackbird, much smaller than the woodpecker with ash-
coloured wings. It frequents even seashores. The red shank,
that is the rubripes,[377] is smaller than both. It has a shorter
beak: but is shorter with red legs (from which the Irish make
up the name). The legs are long, its colour is pale yellow. It visits
mostly the banks of rivers.

*Snipe*      The snipe is the size of a starling with longish legs and beak; its
colour is variegated, with black and pale yellow lines; it lives in
marshes.[378]

# C. XXV
## The Birds less common in other regions

Other birds, which I have neither seen recorded by Latin
authors nor seen elsewhere live in Ireland: of whom some
should be mentioned.

*Black grouse*      The black grouse (capercaillie), equal to a guinea hen in size,
light black in colour is fleshy and not unpleasant to eat. They
frequent the dense wood and they are so astonished at men that
when roosting in a tree they look on while a traveller throws
eight or nine stones. Often they do not fly away until they are
cast down dead. For this reason, I think they are not unsuitably
called stupid birds.

---

376. The redshank between whitish and green in colour frequents rivers
mostly in very cold weather. Larger than the blackbird much smaller
than the magpie, ash coloured wings it frequents also the sea shore.
377. Meaning 'red-footed'.
378. The section was crossed out in the manuscript.

(Dubh ian and Stupida deleted) Tetrax totidem etiam spectant iactus bombardi et ita si occisis aliis et aliae spectandae occidantur. Quamobrem eas non incongruenter stupidas vocari reor.

*Coin defessa*
*chenalopex*

(Aliae anseri magnitudine, coloreque non nihil similes urgente hyemis rigore ex septentrionalibus regionibus per Scotiam Albionensem ad gratam Iberniae temperationem confugiunt. In Tirconella Odonelli principis ditione in spatioso littore post longum volatum plurimae[379] subsidunt, usque adeo fatigatae, ut facile occidantur: ob quod defessae non male nominantur).[380]

*Liatraisc Caerulea*[381]
*vide Arist. de*
*Caerulea L. 9 C.21*
*Hist.*[382]

Caerulea nomen a colore habens, columba minor, nive maxime rigente conspicitur, et moritur.

*Glasain cuilinn*
*Duirridini*[383]
*Fiodhrinn Giodhrinn*
*Runc*

Alia, quam Iberni viridem Paliurum vocant, colore caerulea, carduele minor, decem circiter pullos educit.

Ultimo loco Durridinas[384] aves ortu mirabiles recensebo: pineis lignis[385] in mari diu fluitantibus natae parvae conchae, vel testae haerent.

In his conchis[386] aviculae creantur[387] donec[388] in eam magnitudinem, et firmitatem, quae nando, vel volando idonea sit, crescant. Anseribus similes, sed eis minores in Ibernia frequentissimae conspiciuntur.

Haec de avibus Iberniae scripsisse sufficit, omnes enim comprehendere, non sum ausus profiteri.

---

379. Before 'plurimae' 'fatigatae' is crossed out.
380. The paragraph in brackets is crossed out.
381. Turdus is deleted.
382. This is written in a different hand.
383. Viridis palinurus is crossed out.
384. 'Barneculae' and 'ansermarini' crossed out below Duirridini.
385. 'Sive barnaculas' crossed out after 'durridinas'.
386. 'Quibus' is crossed out and his conchis inserted above the line.
387. 'Formatae' crossed out and 'creantur' inserted above the line.
388. Before 'doner' ex lignis rostro pendent is crossed out.

They often look on at as many shots of a gun as are fired and thus, when some have been killed, others are also killed because they wait.[389]

*Brent goose*

Other birds the size of a goose and somewhat similar in colour, driven on by the severity of winter, flee from northern regions through Scotland to the pleasing moderate temperature of Ireland. In Tirconnel, in a wide beach in the land of Prince O'Donnell, after a long flight very many set down, so fatigued that they are easily killed on account of which they are quite well named 'exhausted.'[390]

*Fieldfare*
*See Aristotle Bk 9, Ch 21 Hist concerning the caeruleus.*

The fieldfare has its name from its colour. Smaller than a pigeon, it is seen in the most severe snow and it dies.

*Green finch*

Another which the Irish call the green Paliurum, dark green in colour, smaller than the goldfinch, hatches out about ten chicks.

*Barnacle goose*

In the final place I shall tell about the barnacle geese, wonderful in their origin.[391] The tiny offspring of a mussel or shellfish cling to pine logs that are tossed about in the sea for a long time. In these shells, little birds are formed until they grow to such a size and strength that they may be fit for swimming or flying. Similar to wild geese but smaller than them, they are very frequently seen in Ireland.

That is enough written about the birds of Ireland: I have not dared to claim that I have included them all.

---

389.   This marginal note is in a different hand.
390.   'on account . . . exhausted' is crossed out in the manuscript.
391.   This story is told by Giraldus Cambrensis, *Topographia*, 1, 11.

# C. XXVI
## Iberniae insecta

Ad avium contemplationem insectorum reptatio proxime accedit: haec iis animantibus, quae retulimus, et terrestribus, et volatilibus corporis quidem robore, structuraque ut plurimum[392] inferiora: immensa tamen artificii spectatissimi subtilitate nonnulla[393] superiora videntur, minime leve argumentum, quam *L. Apes,* incomprehensibili mentis acumine eorum opifex Deus optimus *G. Melissa* maximus polleat! In magnis enim corporibus haud adeo difficilis *H. Abeja* membrorum distributio, materia sufficiente; at in parvis, aut *I. Beach* nullis (ut ita dicam) corpusculis, sicut ape, culice, formica, et minoribus quae scientiae vis, quae rationis acies, quae fabricandi peritia potuit tot membra disponere, tot sensus collocare, tam mirificae industriae instinctum condere? Infinita.

*Virg. 4 Georg*     *'In tenui labor; at tenuis non Gloria'.*
Insecta a sectionibus, vel incisuris, quae cervicum, et aliarum iuncturarum loco membra separant, nomen acceperunt. Non enim nervos, ossa, spinas, cartilaginem, pinguedinem, carnem, crustam, vel veram cutem cum distinctione, sed tenuia massae sibi unique fere similis corpuscula habere dicuntur. Ita divulsis[394] eorum partibus praecipuis (si caput excipias) vivacitas, atque palpitatio aliquandiu inest: quia in his vitales operationes non a certo tantum loco in reliquum corpus, ut in perfectis animalibus, derivantur; sed in singulis partibus principium, et originem ponunt Insectarum altera terrestria altera volatilia sunt.[395]

---

392.   'Ut plurium' is inserted from the R. margin. 'Apes' etc. is crossed out in the margin and re-inserted on p. 152.
393.   'Nonnulla' is added in L. margin.
394.   Illegible word crossed out after 'divulsis'.
395.   'Insectarum altera terrestria altera volatilia sunt' inserted from the R. margin.

# C. XXVI
## Insects of Ireland

The next consideration that comes after the survey of birds is that of insects. These, for the most part, are inferior in strength and structure of body to the beings we have surveyed whether of land or of the air. However, because of the immense fineness of their notable construction, some appear superior. They present a strong proof of how God their creator is mighty, as best and greatest by the incomprehensible sharpness of his intellect. For in large bodies the distribution of membranes is not that difficult when one has sufficient material; but in tiny or (as I may say) non-existent little bodies such as the bee, the fly, the ant and smaller beings, what power of science, what sharpness of reason, what skill in fabrication was able to arrange so many limbs, to assign so many senses, to instill the instinct of such wondrous industry? It is infinite.

*Virg. 4 Georg*     '*The toil is in the tiny thing but the glory is not slight.*'
Insects get their name from sectioning, or incisions which separate the parts of the body at the site of the neck and of other joints. For they are said not to have sinews, bones, spines, cartilage, fat, flesh, shell (rind) or true skin as different items but slight bodies of matter which is almost the same throughout. Thus, when the particular parts are pulled off (if you except the head), the vital force and trembling persists for some time: because in these insects the vital workings are derived not only from a certain spot into the rest of the body as in complete animals, but they place the beginning and the origin in separate parts. Some insects live on the ground, some are flying insects.

# C. XXVII
## Iberniae Apes

Inter utraque[396] communes apes principatum obtinent: quae *1. L. Apes*
non modo mel dulcissimum, atque saluberrimum, ceramque *G. μέλισσα*[397]
hominibus praebent; sed eos etiam ad industriam, militarem *H. Abeja*
disciplinam, reique publicae recte administrandae rationem *I. Beach*
docent. Cum licent illis per tempus ad opera exire, nullum
diem ocio perire sinunt. Favos incredibili ingenio construunt:
et adversus aliarum bestiolarum aviditates muniunt. Primum
tectorium et fundamentum quod eis adhibent, authores[398]
commosin, gummitionem, vel melliginem vocant, amaram,
et glutinosam rem, et ex gummi, resina, lachrymisque salicis,
ulmi, arundinis colligi, docent. Super hoc venit secundum
fundamentum pissoceros dilutior, et ex mitiore gummi
confectus: tertium propolis crassioris materiae qua omnis
frigoris et iniuriae additus obstruuntur.[399] Apes ceras ex
omnium (paucis demptis) arborum, satorumque floribus
confingunt. E cera struunt favos a concameratione alvei orsae.

Favorum cellas uno, vel altero ad summum die melle replent:
quod prima aurora in arborum frondibus, plantarum foliis, et
aliis pabulis dulci liquore roscidis colligunt: haustum in utriculos
suos dimittunt et in alveum reversae ore vomunt: quod Virgilius
*4. Georg.*    insinuavit: *'Protinus aerii mellis caelestia dona Exequor.'*

Plinius plane testatur: et ego comperi: qui ex interfectarum
apum corpusculis hos utriculos molle plenos extrahi vidi.
Dum mel suppetit, eo,[400] cum deest, cera vivunt: cum utroque

---

396.   'Que'? is crossed out at the end of inter and utraque is inserted above
       the line.
397.   G. 'Melissa' is crossed out and 'μέλισσα' inserted next to it.
398.   'Authores' inserted above the line.
399.   'Quadomnis' to 'obstruntur' is inserted from L. margin.
400.   'Eo' is inserted above the line.

# C. XXVII
## Bees of Ireland

Among the two types, common bees have the first place: they *Bee* not only provide the sweetest and most health-giving honey, and wax for men, but they also train them towards industry and military discipline, and the correct conduct of the affairs of state. When the weather permits them to go out to work, they do not allow any day to pass by in idleness. The bees build honeycombs with incredible cleverness, and they defend them against the greedy attacks of other little beasts. The first covering and foundation which they apply to honecomb, authors call commosis, or a gummy substance, or bees' resin. They tell us that there is collected a bitter glutinous substance both from gum-resin and from the weepings of the willow, elm and reed. On top of this comes a second foundation, the pissoceros (the second foundation of the honeycomb, Pl. 11, 6 & 7), which is thinner and made up of a softer gum: the third structure is the propolis of a thicker substance by which all approaches of cold and damage are excluded. Bees make waxes from the flowers of all trees and plants, with few exceptions. From the wax they build the honeycomb from a series of chambers forming the beginning of a hive.

They fill up the cells of the honeycomb with honey to the top in one or two days; which at first dawn they collect from the foliage of trees and the leaves of plants and from other foods dewey with a sweet liquid: they inject what they have sucked up into their little stomachs and vomit it from the mouth on their return to the hive: which Vergil suggested:

*Georg. 4,1 & 2* *Straight away I will describe the celestial gifts of honey from the sky.*[401]

Pliny clearly testifies and I have discovered because I have seen those little stomach sacs full of honey being extracted from the little bodies of bees who have been killed. While the honey

---

401.   The ancients considered that honey fell with dew from heaven.

inter componendos favos carent, alium sibi cibum comparant ex rore, et arborum succo gummi modo genitum, qui saporis amari erithace, sandaracha, et cerinthus vocatus in favorum inanitatibus sepositus quandoque invenitur. Mellificant vero in Ibernia non solum in alveis ad id ab hominibus factis; sed etiam in arborum cavis, sepium rimis, terrae foveis; unde sequitur mellis, ceramque proventus uberrimus. Hanc regendi rationem servant. Interdiu stationem ad portas more castrorum collocant. Tenebris noctis approprinquantibus in alveo strepunt, donec una gemino, sive triplici bombo, sicut buccino quietem capere imperet. Tunc repente conticescentes in sequentem lucem quiescunt, donec simili signo excitentur. Inde ad opera provolant, si dies mitis futurus est: namque ventos imbresque praedivinantes cavent: Noctu in expeditione deprehensae supinae excubant, ut alas a rore protegant, quominus volare prohibeantur. Adverso vento tendentes sumptis lapillis sese firmius librant, ut potius progredi possint, quam recedere cogantur.

*4. Georg.*

Cuius rei mominit Maro:
*Excursusque breves tentant, et saepe lapillos*
*Ut cymbae instabiles, fluctu iactante, saburram*
*Tollunt: his se per inania nubila librant*

Segnium inertiam notant, mox castigant, denique morte puniunt. Eademque latrocinium plectunt. Aculeo non, nisi omino diro suo, foriunt. Unde cecinit quidam:[402]
*Hoc liquet ex apibus, quae levia spicula figunt:*
*Mors tamen tunc ipsis accelerata venit*

---

402.   Illa e? brevis in R. margin.

lasts, they live on it; when it is lacking they live on wax: when both are lacking during the building of the comb they provide another food for themselves from the dew and from the juice of gum off trees that has just been produced which is called bee bread of bitter taste, or sandracha, or corinthus, and is put aside in the empty space of the combs whenever it is found. Indeed in Ireland bees make honey not only in hives, made by men, for this purpose, but also in cavities of trees, in fissures in fences, and in holes in the ground, from which follows the most abundant produce of honey and wax. They observe this system of ruling. By day, they place a guard at the gates after the manner of military camps. When the darkest of night is approaching, they make a loud noise in the hive until one bee with a two-fold or three-fold buzz as if with a blast of a trumpet gives the order to sleep. Then suddenly becoming silent, they stay quiet into the following morning until they are aroused with a similar signal. Then they fly to their tasks if the day is about to be mild: for they foresee winds and showers and shun them.[403] If they are caught out by nightfall on an expedition they sleep out of doors on their backs in order to protect their wings from the dew, lest they be prevented from flying. When they are venturing against an adverse wind, they take up little stones and they balance themselves more firmly so that they can progress rather than be compelled to retreat. Which event is recorded by Virgil.

*Georgics 4*

'They attempt short journeys and often they carry pebbles just as unstable skiffs, thrown around by the wave carry sand as ballast. With these they balance themselves through the empty clouds. They note the inactivity of the lazy ones, soon they reprove them and then they punish them with death.[404] And in the same way they punish robbery with blows. They don't strike with the sting except in an extreme threat to themselves.' From this someone has sung:
*This fact is clear from bees who affix the light stings,*
*However, swift death then comes to them themselves.*[405]

---

403.   The correct reference is Pliny 11,28,113.
404.   Morte was left out by O'Donnell in transcription.
405.   The origin of this quotation is not clear.

Reges (alia species apum est illa) habent, quibus an sit aculeus? *2. L. rex apum*
non constat: sed eo illos minime pungere inter authores convenit. *G.*
Illis in alveo regias ampliores, atque magnificentiores, quam *H. Maestro dela*
sibi aedificant, mel parant, eos alveo egressos frequenti agmine *colmena*
stipant, summe venerantur, quod et Virgilius animadvertit. *Ib. Ri na mbeach*
'Praeterea regem non sic aegyptus, et ingens *Beach gabhair*
Lydia, nec populi Parthorum, aut medus Hydaspes Observant*[406]* *Vide vespa*

*Tabanus idem asilus* Inter duos reges cum de principatu lis vertitur eam acri
*Pl. L11. C.28* cortamine apes dirimunt, attestante Marone:
*An Creabhair keach* *Tum intrepide inter se coeunt, pennisque coruscant*
*Spiculaque exacuunt rostris, aptantque lacertos:*
*Et circa regem, atque ipsa ad praetoria densae*
*Miscentur, magnisque vocant clamoribus hostem*

Fucis quoque aculeo minime armatis, alii speciei e suo genere *3. L. Fucus*
mella largiuntur et domicilia in alveo construunt, sed suis *G. κηφήν[407]*
minora. *H. Zangano*
*I. Ladran Shtathi*

## C. XXVIII
### Formicae Iberniae

*4. L. Formica* Ab apum industria, providentiaque minime abhorrent
*G. μύρμηξ[408]* formicae,[409] a ferendis micis vocatae. Est autem mica proprie
*H. Hormiga* spectata pulvisculus, qui in arena, quasi argentum fulget, a
*I. Siangain* quo panis aliarumque rerum minima frustula nomen idem
sumpserunt. Formicas duntaxat communes, alisque destitutas
gracilia quidem, et exigua corpuscula, animi tamen[410] plena
Ibernia progignit. Quae cibos sibi[411] mitibus diebus comparantes

---

406. After 'observant' the author continued to write 'inter duos reges cum
de', but crosses it out and transfers it to the next line.
407. G. Cephne is crossed out in R. margin and κηφήν inserted from the L.
margin.
408. G. Myrmex' crossed out and 'μύρμηξ' inserted below it.
409. After 'formiace' illegible word crossed out.
410. 'Ingentis' deleted after 'tamen'.
411. After 'sibi' 'quamis non ut apes faciunt' plus two illegible words are
crossed out and 'metibus' is inserted above the line.

They have Queens[412] (this is another species of bee): I do not *Queen bee* know if they have a sting, it is not agreed. It is agreed among authors that they do not sting with it. The bees build for the queens, in the hive, larger and more magnificent palaces than for themselves. They prepare honey, they surround them as they go out from the hive in a line of dense numbers; they respect them highly, a fact that Virgil also noticed.

*Besides, neither Egypt nor great Lydia nor the people of*
*Parthia nor Persian Hydaspes honour thus their king.*

When a contest arises between two Queens concerning who rules, the bees separate in a bitter battle, as Maro attests:
*Then fearlessly they come together, they flash their wings, they*
*sharpen their stings with their beaks and they get ready their arms.*
*Around the queen and around the cell of the queen bee they gather*
*in swarms and call on the enemy with mighty noise.*

To the drones, another species of their own kind not armed *Drone* with a sting, the bees give a gift of honey and they construct dwellings for them in the hive but smaller ones than their own.

*irg Georgics 4,*
*3–76. Gadfly*
*nd horsefly are the*
*me Pl. L. 11, 28*
*reabhair Keach*

# C. XXVIII
## The Ants of Ireland

*nt*[413]

Ants are akin to bees in industry and foresight; they are called formica because they carry grains.[414] Mica, properly speaking, is a fine dust which shines like silver in the sand; from which the tiny crumbs of bread and of other materials have taken the same name. Ireland produces at least common ants devoid of wings. Their bodies are slender and tiny but full of mighty courage.

---

412. The queens are referred to as *reges* (kings) in the text because the ancients did not realize they were female.
413. Pl. 11,36,108.
414. Mica = a grain.

in aspera tempora servant, terra conditos. Semina non, nisi arrosa condunt, ne rursus a terra in fruges exeant: pondera minora morsu gerunt, maiora aversae postremis pedibus moliuntur: quae formainis introitus non capit, dividunt: imbre madefacta ne corrumpantur, ad solem, ventumque prolata siccant. Noctu quoque plena luna operantur: interlunio cessant. Quoniam ex diverso victum convehunt aliae aliarum ignarae, certis diebus ad mutuam recognitionem conveniunt.

Inter se sepeliri memoriae proditum est.

## C. XXIX
### Araneus, culex

*S. L. Araneus*  
*G. ἀράχνη*[415]  
*H. Arana*  
*I. Dubhain Fhalla*

His araneorum contemplationem non absurde iungemus: ex eorum[416] generibus innocuos illos Ibernia gignit: qui artificio mirifico stamina[417] collocant, subtegmina, vel tramam annectunt, et inde telas quibus, velut retibus bestiolas capiunt, texunt. Illi vero teretes, atque subtilissimi fili,[418] araneam componentes qua ex materia fiant, dubium fuit, eos antiquorum Philosophorum aliis ex araneorum excremento, aliis ex tenui membranula, quae velut cortex araneum cingat, nexi putantibus.

---

415.　Mygala crossed out and 'ἀράχνη' inserted next to it.
416.　'E quorum' is crossed out and 'ex eorum' inserted above the line.
417.　'Tela' crossed out before 'stamina'.
418.　After 'fili' 'quibus' is crossed out and after 'araneam' 'confiutur' is deleted.

They save the foods which they collect during the mild days for the hard times, having concealed them in the ground. They do not hide seeds unless they have been gnawed lest they put out shoots again from the earth. Smaller weights they carry in their mouth, the larger ones they push with their back legs when they have turned round. What the opening of the doorway does not take they divide, lest seeds, when softened by a shower, be spoiled, they dry them by taking them out into the sun and the wind. They work during the night when the moon is full; they cease during the time of the new moon. Since some bring food from different areas, unknown to each other, on certain days they get together for the purpose of exchange of information. It is recorded that they are buried among themselves.

## C. XXIX
### Spider, Fly

*ider*

After these we shall reasonably adjoin an examination of spiders. Out of the breeds of spider, Ireland produces only those which are harmless. With wonderful craftsmanship they lay down the warp and weave in the weft or the woof and thus they spin webs with which they capture little creatures as if with nets. Truly, there has been a doubt about what material those smooth very fine threads which make up the cobweb are made from. Some of the ancient philosophers considered that they were made from the excrement of the spiders, others that they were woven from a thin membrane which surrounds the spider like skin.

6. *L. Culex*
*G. κώνωψ*[419]
*H. Mosquito*
*I. coirrimhil*

Culicum hic duo notissima genera conspiciuntur: utrumque volatile, utrumque molestum, utrumque tamen parvum; sed alterum maius: quod non modo hominum cutem; sed etiam equorum, et aliorum animantium tergora acutissimo spiculo fodiens, sanguinem, nisi prohibeantur, suggit. Minores aestate sub vesperum circumvolitando facies hominibus obruendo, non parum fastidii pariunt.

# C. XXX
## Scarabaeus, cicada, gryllus, papilio musca

Utrisque multo maiores scarabaei hic cernuntur, quibus binae fragiles pennae sunt inditae: Pennarum vero fragilitatem, atque tenuitatem durior superveniens crusta[421] munit.

7. *L. Scabareus*
*G. κάνθαρος*[420]
*H. Escarabajo*
*I. Skeartain crubach*

Cicada inde dicta, quod cito cadat: nam aestivis duntaxat mensibus apparet, cognoscitur cantus sui garrulitate, cuius meminit Virgilius, *'et cantu querulae rumpent arbustae cicadae'* Membraneis pennis[423] incedit: rore vivere refertur.

*4. Georg.*

8. *L. Cicada*
*G. τέττιξ*[422]
*H. Cigarra Chicarr*

*L. Gryllus*[424]
*G. γρύλλος*[425]
*H. Grillo*
*I. Dreollin Teaspa*

(Huic similis gryllus, longiore corpusculo retro saltat, terram terebrat, noctu stridet.)[426]

Papillions Iberniae, quibus cornicula ante oculos praetenduntur, mire nascuntur, atque moriuntur. Eruca vermis, qui in olerum foliis immobilis accrescit, inde tecta duro cortice, et araneis

9. *L. Papilio*
*G. ἤψυκος σκοτοδοκ*
*πυραυστής*
*H. Maripossa*
*I. feileacain*
*(féileachán)*

---

419. G. 'Conops' crossed out and κώνωψ inserted after it. Illegible letters cross out at the end of 'fudiens' and 'suggit'.
420. G. Cantharos crossed out and inserted 'κάνθαρος' above it.
421. 'Munit' is crossed out and reinstated.
422. G. Tetix crossed out and G. τέττιξ inserted.
423. After 'pennis' 6–7 illegible words are crossed out.
424. G. Epsieche, Pyralis, Pyrusta, I. Feilarchan crossed out.
425. G. Grylus crossed out and 'γρύλλος' inserted next to it.
426. The words in brackets are inserted from the L. margin.

*Midge*  Two well-recognised types of midges are seen here; both are winged, both annoying, and both however are small. But one of these is bigger: this type piercing not only the skin of men but also of horses and other animals with a very sharp sting suck the blood unless it is prevented. The smaller ones in the summer, towards the evening time, by flitting around and covering the faces of men, cause no little disgust.

# C. XXX
## The tick, cricket, grasshopper, butterfly, fly

Beetles much larger than both midges are seen here. They have *Beetle* two sets of fragile wings. A harder crust on top protects the fragility and fineness of the wings.

The tree cricket gets its name because it falls quickly, it appears *Cricket*[427] mostly in the warm months; it is recognised by the chattering of its song, about which Virgil said and
*The querulous tree crickets shall burst forth with song.*[428]
It travels on membranous wings; it is said to live on the dew.

*Grasshopper*  The grasshopper is similar to the cricket. It has a larger body and jumps backwards. It bores through the ground and it makes a harsh noise at night.

The butterflies of Ireland, before whose eyes little horns stretch *Butterfly* out, are born and also die in a wonderful way. The caterpillar worm which increases in size, immobile, on the leaves of vegetables, and then covered by a hard covering and protected from spiders is called a chrysalis. From that finally, when the covering bursts, out flies the butterfly, which is of such a nature

---

427. Plin. x1, 32.
428. Virg. G. 3,328; E. 2,13. The quotation is different from the modern standard text.

obsita chrysalis appellatur. Ex ea denique, cortice rupto, papilio evolat; qui ea est natura, ut circum lucernas volitans ambustis alis pereat: Unde Pyraustae gaudium gaudere ii proverbialiter dicuntur, qui momentanea, fluxaque voluptate delectati, temporis, corporis et (quod summum damnum est) animae iacturam faciunt. Scio tamen pyraustam, vel pyralem etiam vocari Plinio pennatum quadrupes quod maiori muscae par in Cypri aerariis fornacibus genitum, in igne vivit, et cum ab eo paulo longius procedit, perit.[429]

Non est quid latius muscae naturam prosequar animalis inter nos versatissimi, atque adeo in Ibernia notissimi bipennis, quod cibi avidissimum a ferculis vix, et muscario flabello ad id constituto abigitur.

*10. L. Musca*
*G. μυῖα*[430]
*H. Musca*
*I. Cuil*

Unde quidam cecinit non illepide
*Musca, canis, mimi sunt ad convivia primi*
*Non invitati veniunt prandere parati*

## C. XXXI
### Ricinus, hirudo

*11. L. Ricinus*
*G. κρότων*[431]
*H. Rezno garrapata*
*I. Skertain*

Ricinus quoque non minus notus, quam molestus est: qui infixo semper capite, sanguine vivit: atque ita cum illi cibi non sit exitus, nimia satietate dehiscens, alimento moritur: vivus canes, boves, et alia etiam animalia acerrime infestat.

*12. L. Hirudo*
*sanguisuga*
*G. βδέλλα*[432]
*H. Sanguisuela*
*I. Sumudoir*

Non dissimilis omnino sitis est hirudini in Ibernia frequenti, quae sponte non, nisi plena, sanguinem, quem sugere coepit, relinquit: quod et Horatius memorat:
*Non missura cutem, nisi plena cruoris hirudo*
ea longula, et aquatilis est.

*In arte poetica*

---

429.    Inserted in a different hand in R. margin: Pyraustae gaudium vide? Heraimur in? choil? Crossed out.

430.    G. Myia crossed out.

431.    G. Croton is crossed out.

432.    G. Bdella crossed out and βδέλλα inserted.

that it perishes by burning its wings flying round oil lamps. From this, those people who take delight in the changing pleasure of the moment and cast away their body and (the worst loss) their soul.[433] I, however, know that a winged quadruped equal in size to a larger fly, is called by Pliny[434] *pyrausta* or *pyralis*; it is born in the copper smelting furnaces of Cyprus; it lives in fire and when it goes away from it a little too far it dies.

There is no need why I should pursue the nature of the fly any further, a creature most common among us and so in Ireland most well known. It has two wings, and it is most avid for food mostly from dishes and is scarcely kept away from dishes, even with a fly whisk made for that specific purpose. From this someone has written, not without wit:
*The fly, the dog, and actors are first to the feast.*
*They come uninvited, prepared to eat.*

## C. XXXI
### The tick, the leach

The tick also is as well known as it is troublesome; with its head always stuck in, it lives on blood: and thus since it has no exit for that food, bursting with too much satiety it dies from nourishment. While alive, it grievously infests dogs and cattle and also other animals.

The leech has a not at all dissimilar thirst: it is common in Ireland. Unless full, it does not of its own accord abandon blood which it has begun to suck up: a fact also that Horace records:[435]
*'The leech is not about to give up the skin unless full of blood.'*
It is quite long and lives in water.

---

433.   Pyraustae – winged insects who live in fire.
434.   Plin. 10, 42.
435.   Horace *Ars Poetica* 476.

## C. XXXII
### Teredo, blatta, lumbricus, taenia

*13. L. Teredo*
*G. τερηδών*[436]
*H. Gusano que rue la*
*madera*
*I. milcrinn*

Mira sed et damnosa est teredo, quae cum sit vermis parvus, robora tamen, et alia ligna rodit, cariem inducens; ob quod minime potest Ibernis non cognosci.

*14. L. Tinea*
*G. σής*[437]
*H. Polilla*
*I. Ileain*

Haud omnino dissimili malificio sese tinea manifestat, vermiculus, qui situ, pulvere, vel putredine nascens, vestes, atque libros rodit.[438]

*L. Blatta*
*H. Cucurracha*
*I. Kiarog*

Blatta nomine gaudet vermis qui alveria laedit et alius in balneis prognatus quem blattam lucifugam dicunt.

*15. L. Tinea*
*lumbriscus, Tinea*
*G. Ἕλμις*[439]
*H. lombris*
*I. piast agulsi speain*

Est et aliud tinearum genus, quas etiam lumbricos et taenias vocant. Has graciles, et longulas tum in ipsa terra, tum in hominum corporibus in Ibernia (sicut et alibi) nasci, comperimus.

---

436.  G. Teredon crossed out and τερηδών inserted.
437.  After tinea lumbricus is crossed out. Two illegible words crossed out, one before and one after σής. An illegible word is crossed out under Polilla, I. Illean inserted above, two deleted words 'piast agulsi' and 'spenian'.
438.  After 'rodit' the following is crossed out. 'Hanc blattam qua lucifuga? nominatur nonnulli vocarunt: quo etiam' above 'quo etiam' blatta is inserted above the line.
439.  G. Helminthos is crossed out 'Ἕλμις' and inserted underneath.

# C. XXXII
## Woodworm, Clothes moth, Earthworm, Tapeworm

*Woodworm*

The woodworm is amazing but also damaging, although it is a small worm, it erodes oak trees and other woods bringing on decay; for this reason it is impossible for the Irish not to know it.

*Moth*

In causing quite similar damage, the moth shows itself. As a little worm being born in dirt, dust, or in decaying matter, it gnaws at clothes and books.

*Chafer see Blatta.*[440]

A pest that damages beehives rejoices in the name blatta (chafer). There is another sprung from humid vapour in baths, which they call the light fleeing chafer.[441]

*Tapeworm*

There is also another family of worms which are called maw-worms or tapeworms. I have found that these, rather thin and long, are born both in the ground itself and in the bodies of men in Ireland (just as elsewhere also).

---

440.   Written in right margin.
441.   This is an insertion from the R. margin.

# C. XXXIII
## Lacerta, Cochlea, acarus[441]

(Lacertae hic passim conspiciuntur; sed sine veneno, et omnino *L. Lacerta*
innocuae, nisi dormientis sub dio os fortassis ingrediantur: quod *G. σαύρα*[442]
vix semel vel bis universis saeculis contigisse accepi.)[443] *H. Largarto*
*I. Airkluachra*

Cochleae domiportae, et carentes testa limaces utrique humi *L. Cochlea*
serpentes adiungantur. *G. κοχλίας*[444]
*H. Caracol*
*I. Selihidi*[445]

Non est quid alia insecta memorem, quae Ibernia multa gignit, *L. Acarus*[446]
usque ad acarum animalium minimum, qui in viventis hominis *G. ἀκαρής*
carne generatur. *H. Arador*
*I. Frith*

# C. XXXIV
## Quibus animalibus caret Ibernia

At cimicem, vermiculum foedum odore, mascentem ex ligno;
dormientes infestantem, ranam, scorpionem, bufonem,
viperam, colubrum aut ullum venanatum animal sive insectum,
sive perfectum haec insula minime procreat, ut suo loco

---

441. The following is crossed out in the text: 'Non est quid alia insecta
memorem quae Ibernia multa gignit usque ad acarum animalium
minimum quiin vivintis hominis carne gignitur. At cimicem
vermiculum foedum odore nascentem ex ligno dormientesque (?)
infestatem ranam scorpionem, bufunem (ranam serpentem crossed
out). Viperam, colubrum aut ullum venenatum animal sive insectum
sive perfectum haec insula minime procreat, ut suo loco fustus
dicemus. Nam nunc vereor ne tam longa cuolatilium crossed out and
terrestrium volatiliumque narratione lectori taedium pariam, quo
metu ut eum imposterum liberum aquatileum reputationem breviter
percurrere conabor'.

442. G. saura crossed out and σαύρα inserted.

443. Crossed out in the L. margin, L. Cochlea, H. Caracol, I. Shelidi. Below
this some illegible words are crossed out.

444. G. Cochlias crossed out and κοχλίας inserted.

445. In R. margin abvoe lacerta L. Acarus, G. Alarus, H. Aradur, I. Frui are
crossed out.

446. G. Acaros crossed out and ἀκαρής inserted.

## C. XXXIII
### Lizard, snail, flesh worm

Lizards are seen everywhere here; but without poison and *Lizard*
totally harmless except that they may enter the mouth of one
sleeping in the open air: a fact that I have heard to happen
scarcely once or twice in all ages.

The shell-bearing snails and slugs without a shell, both creeping *Snail*
on the ground, may be added.

There is no reason I should mention other insects, which Ireland *Flesh worm*
produces in numbers, down to the flesh worm, the smallest of
animals which is born in the flesh of a living man.

## C. XXXIV
### Animals not found in Ireland

This island does not give birth to the cimex, a little worm with
a foul smell, born from wood, which infects sleeping people.
This island does not produce the frog, the scorpion, the toad,
the viper, the serpent or any poisonous animal or insect as we
shall discuss in its proper place.[447]

---

447.  At this point in the manuscript, a section has been erased, and the
      material has then been incorporated at the start of the subsequent
      section.

dicemus. (Crossed out: nunc vereor ne tam longa terrestrium, volatilium que narratione lectori taedium pariam, quo metu ut eum imposterum liberim aquitilium reputationem breviter recurrere conabor).

## C. XXXV
### Iberniae Aquatilia

Equidem vereor, ne tam longa terrestrium, volatiliumque narratione lectori toedium pariam. Quo metu ut eum imposterum liberem, aquatilium reputationem breviter percurrere conabor.

## C. XXXVI
### Balaena, equus, et canis aquaticus, vitulus marinus, Delphinus

Principio quaedam vasta magnitudine balaenae, animantia ex genere cetorum[449] quandoque in Iberniam appellunt: Inter quarum immanium beluarum species orca physeter,[450] Pristes rota numerantur.

*1. Latine balaena*[448]
*G. φάλαινα*
*Hispanice Vallena*
*Ibernice Milmór*

Ibi conspiciuntur alia aquatilia, corio, pileque intecta, ut equi aquatici, perraro visi magnorum lacuum incolae, terrestribus similes sunt[451] hyppopotami. Id usque adeo verum est, ut inter equos pascentes nonnunquam minime dignoscantur.

*2. L. Equus aquiticus*
*G.*
*H.*
*I. Each Iski*
*Locheachoir*
*Maccabhuil rapitur*
*in hunc lacum eius*
*pedissequus, qui lacus*
*est in tirona, et ipse*
*hippotaumus erat*
*taurea colore, unde*
*lacus nomen accipt.*

---

448. Graeie Bhalaenac crossed out.
449. Animalia ex genere cetorum is crossed out after cetorum
450. At the end of 'physeter' the 'seter' is crossed out along with pristes? and rota? and the 'seter' is transferred to the next line.
451. 'Maiores multo quam' crossed out and 'similes sunt' inserted above the line.

# C. XXXV
## Aquatic creatures of Ireland

Indeed, I fear that I might induce tedium in the reader by such a long account of terrestrial and flying animals. So that I may, for the following, free him from that fear of deceivers I will try to run through briefly the consideration of aquatic creatures.

# C. XXXVI
## The whale, sea horse, otter, seal, and dolphin

First, whales of vast size, animals from the family of sea monsters, *Whale* land sometimes in Ireland. Among the species of immense beasts, the sperm whale, the killer whale, the Pristes[452] and Rota[453] are numbered.

Other aquatic animals are seen there, covered with hide and *Sea horse* with hair, such as water horses who, as inhabitants of great lakes, are rarely seen. River horses are similar to land horses. That fact is so true that they sometimes are not distinguished when feeding among horses.

---

452.   Pristis is any sea monster or whale. Translated by H. Rackham as a shark in Pliny 9.3.8., in Virg *Aen.* III, 427, translated as sea monster Loeb Class. Lib. edited by T.E. Page, ECAPPS, W.H.D. Rouse, La Posit and E.H. Warmington.

453.   Rota, a kind of sea fish. Pl. 9.4. §8. Pliny says they are called wheels because they simulate cart wheels with four radiating spokes, with the two eyes located on each side of the hub.

Obuilli viri inter Ibernos satis noti casus infaelis fidem pandit. Hic hyppopotamum,[454] qui inter terrestres pascebatur, phaleris instructum sinistro omine ascendit. Hippopotamus[455] proximum lacum praeteriens in illum sese cum assessore deiecit, nec amplius ipse apparuit, nec obuillus extitit. In rei memoriam lacus obuillus nomen accepit.

*Loich Ibhuill*

*3. L. Canis Aquaticus* Aquatici canes lutra[456] maiores memorantur aliquorum
*G.*            fluminum habitatores; qui urinantes, atque natantes,
*H.*            acutissimo[457] rostro (nisi res fabulosa fertur) quandoque
*I. Anchu nutriae*    transfodiunt. Lutrae vero naturam superius explicatam
              impraesentiarum repetere, supervacaneum duxi.[458]

*4. L. Vitulus marinus* vituli marini[459] circumiacentia Iberniae aequora, et littora
*G. φώκη*[460]      confertis agminibus frequentant.
*Lobo marino I. Roon*

*Bradán Fearna, Bradan ri, sturgeon habet dursum viride ei asperum vide sturio vide silurus. An mero hispanice?*

*5. L. Delphinis*    Eadem Delphini corio tantum vostiti, et amore in homines[461]
*H. Delphin*        celebratissimi solent celebrare.
*G. δελφίν*[462]
*I. Muic mairi*[463]

---

454.  'Aquaticum? equum' crossed out and 'hippopotamum' inserted above the line.
455.  'Hippopotamus' inserted above the line, illegible word crossed out underneath.
456.  'Nutra' crossed out and 'lutra' inserted above the line.
457.  'Qui' crossed out before 'acutissimo'.
458.  'Nutra' is again crossed out and 'lutra' inserted above the line.
459.  After 'marini', 'aquoroum' is crossed out.
460.  G. Phoca crossed out in the L. margin.
461.  After 'homines' an illegible word is deleted.
462.  G. Delphin crossed out in the L. margin.
463.  *Muc mara* is actually a porpoise. (Dinneen)

*Locheachoir The*
*Maccabuil: flows into*
*his lake a tributary*
*of it, this lake is in*
*Tyrone*

The unhappy death of a man of Obuillus, (O'Boyle) quite well-known among the Irish, explains the belief. This man by misfortune mounted a river horse fitted with trappings while it was feeding among the land horses. The river horse travelling by the nearby lake threw himself along with his rider into it. The river horse itself did not appear again nor did O'Boyle survive. The lake received the name Obuillus in memory of this event.

*The water-horse itself*
*was like a bull in*
*colour. From which*
*the lake got is name*
*Loch I Bhuill.*

*Water dog*

Water dogs larger than the otter are recorded as inhabitants of some rivers. They sometimes pierce diving and swimming birds with their very sharp snout (unless the story was made up). Truly, I have decided that it is redundant at present to recall the nature of the otter which we explained above.

*Seal*

Seals frequent the beaches and shores lying around Ireland in dense masses.

*Sturgeon*

*The sturgeon has a hard green back. See sturio and silurus.*[464] *Is this mero in Spanish?*

*Dolphin*

In the same way dolphins are clothed in thick skin and are famous in their affection for humans.

---

464. Silurus (sheath fish), another name for the European catfish.

# C. XXXVII
## Pisces longi

*6. L. Thynnus*
*G. θῦνος*
*H. Atun*
*I. Thuinach*[465]

Piscium vel squamis, vel cortice modo molli, modo aspera indutorum innumera propemodum genera hic reperiuntur: ut thynnus, cuius generi orphus, atque pompilus adscribuntur:

*7. L. Capito*
*G. κέφαλος*[466]
*H.*
*I.*[468]

ut capito a capitis magnitudino nomen habens(quieti labeo dicitur)[467] reliquo etiam corpore non exiguus: quibus et sequentes adnumerentur.

*8. L. Asellús*
*G. ὀνίσκος μάξεινος*[470]
*H. Merluza*
*Zezial, pescada*
*I. Coilm oir. Trosc*
*vide milvus*

Asellus Iberniae[469] capitoni magnitudine fere par, sed eo carne nobilior, et recens, et sole siccatus magnae authoritatis habetur.

Draco marinus captus,[471] si vivus, in arenam immitatur, celeritate mira cavernam sibi rostro excavat.

*9. L. Draco*
*G. Dracon*
*H. δράκων*[472]
*I. Cnudain*
*Macreal vide*
*scombrus, magarus*

Congrus oblongus, atque lubricus, anguillac magnae non dissimilis acerrimas inimicitias cum muraena gerere fertur.

*10. L. Congrus conge*
*G. κόγγρος*
*H. Congrio*
*I. Congrio.*
*Deargan vide rubillio*
*aurata critrinus,*
*tragus pragus et hisp.*
*betogo et belugo*

---

465.   I. Bradan Fearna is crossed out in the L. margin.
466.   G. Kephalos is crossed out in the L. margin.
467.   'Quieti labeo dicitur' is inserted from the R.margin.
468.   ? Truisc crossed out in the L. margin.
469.   Iberniae is inserted above the line.
470.   G. Oniscos deleted in L. margin and ὀνίσκος, μάξεινος are inserted from the R. margin.
471.   'Captus' is inserted above the line, 'si captus' is crossed out and 'si vivus' inserted above it.
472.   H. Dragun deleted and δράκων inserted in the L. margin.

# C. XXXVII
## Long Fish

*Tunny*

Here are found almost innumerable types of fish covered either with scales, or with skin sometimes soft, sometimes hard; such as the tunny to whose family the gilt head and the pilot fish are ascribed,

and such as the capito which gets its name from the size of its head, (it is also called labeo). The rest of its body is not small. With these also are classed the following:

*Hake*

The hake of Ireland is almost equal in size to the capito, but the flesh is much finer; either fresh or dried by the sun, it is much prized.

The sea dragon, when caught, if he is put on the sand alive, *Sea dragon* excavates a hole for himself with his snout with wonderful rapidity.

*Mackerel see scombrus and magarus*

The conger eel is long and slippery, not dissimilar to a large eel. *Conger eel see rubellio,* It is said to have a most bitter enmity with the lamprey. *aurata critrinus tragus, pragus, and Hisp. betugo and belugo.*[473]

---

473. None of these names could be found apart from *tragus* which is a 'kind of fish' Plin. 32,11,54. § 152 and rubellio aurata, a fish of a reddish colour Pl. 32, 10, 49 § 138

Haec eadem fere magnitudino molli cute induta multis maculis, quas oculos vocant, picta foemina est cuius mas myrus vocatur.

*11. L. Muraena*
*G. μύραινα*[474]
*H. Morena Lamprea*
*Ib luimpraea*
*Briendi fluviatris*
*piscis*

Hic quoque capitur torpedo, piscis cartilagineus, aspera cute, nomen sortitus non quod ipse torpeat, sed quia limo mersus pisces secure supernatantes torpefactos corripiat; quam vim mortuus non habet, esu utilis.

*12. L. Torpedo*
*G. νάρκη*[475]
*H. Tremielga*
*I. coilleach breach,*
*gobog lius*
*vide lupus piscis et*
*lucio apud huertam p.*
*551, col. 2*

Ei figura, et cutis asperitate similis est lupus, etiam cartilagineus, sed minor olim in pretio longe maiore, quam hodie habitus. Luporum qui a candore mollitiaque carnis Lanati dicuntur, laudatissimi erant.

*13. L. Lupus*[476]
*G. λάβραξ*
*H. Sollo*
*I. fiagach lanatus*[477]

Eis hodie in Ibernia sicut et alibi mullus multo nobilior est, qui cum geminam barbam ab inferiore labro pendentem habet, barbatus, sive barbatulus // nuncupatur: mugili hostis.

*14. L. Mullus*
*G. τρίγλα*
*H. barud salmonete*
*I. Millet, barbatus*
*muilleat*[478]

*15. L. Mugil*
*G. κόδολα πλώς*
*H.*
*Ib muigil*
*De mugile vide Plin.,*
*calep. et huerta,*
*p. 569 cuius genus*
*fluiatile vocatur, Hisp.*
*Albus et est in Ibernia*

Nam mugil in Iberniae oceano generatur capitosus, omnium piscium squamis intectorum velocissimus, qui in metu capite abscondito se totum occultatum credit.

---

474.   Myraena deleted and μύραινα inserted.
475.   G. Narce deleted and νάρκη inserted.
476.   L. Lupus – pike; G. λάβραξ – Bass. H. Sollo? I. fiagach – dogfish.
477.   After lanatus an illegible word is deleted.
478.   'muilleat' has been crossed out in the manuscript.

The lamprey is almost of the same size covered in a soft skin
with many spots which they call eyes, the female is coloured
and her male partner is called a myrus.

*Lamprey*
*Briendi a river fish*
*(unknown).*

Here also the electric ray is caught, a cartilaginous fish with a
rough skin, it gets its name not because it is sluggish but, because
submerged in mud, it stuns fish which are unconcernedly
swimming above; when dead, it does not have this power; it is
useful to eat.

*Electric ray*
*See Lupus Piscis*
*Lucio in Huerta*
*p551, c01.2.*

The dogfish is similar to the electric ray in shape and in the
roughness of the skin; it is also cartilaginous though smaller;
but in times past it was worth more, by a long way, than it is
today. Of the dogfish those called lanati, from the whiteness
and softness of the flesh, were the most praiseworthy.

*See dogfish and Lucio*
*in Huerta p551,*
*c01.2.*

The red mullet in Ireland today, just as elsewhere, is much finer
than these; since it has a double beard hanging from its lower
lip it is called bearded or a small-bearded. It is an enemy to the
(ordinary) mullet.

*Red mullet*

*Mullet*
*Concerning the*
*mullet see Pliny*
*and Calepinus and*
*Huerta p. 569.*[479]
*The river type in*
*Spanish is called*
*Albus, it is also*
*present in Ireland.*

The mullet in the ocean of Ireland is born with a large head.
It is the speediest of all the fish that are covered in scales, and
when frightened, when its head is covered, it believes that it is
completely hidden.

479. O'Sullivan uses editions of Pliny by Decacampius and Huerta.

Ichthyocolla piscis glutinosum habet corium; et eius glutino idem quoque nomen inditur.

*L. Ichthyocolla*
*G. ἰχθυοκόλλα*[480]
*H. Cazon*
*Ib.*

Halece quae maena, scombrusque vocitatur aqua duntaxat victitante, Ibernia abundat, non magno nobili tamen atque notissimo pisciculo.

*16. L. Halex Halex*
*G. μαίνα, μαίνις*[481]
*H. Harenca, sardina*
*I. scadain, serdin*
*pilser*[482]

Fluviatilium, quibus Iberniae flumina nobilia, lacusque scatent, genera non sunt praetereunda, salmones totius orbis optimi: tructae et anguillae Vide Huertam 180 de truta carpion, et Pion.

*17. L. Salmo*
*G.*
*H. Salmon*
*I. Bradain*

*Vide Plin. de salme Lib. 9. c.18 et aliqurum cibum piscibus, omnibus praeferend.*

*18. L. Tructa*
*G.*
*H. Trucha*

19. L. Anguilla G. ἔγχελλος ἔγχελις[483] H. anguila I. ascu

*I. Breach de trocta*
*Truta vario de tructa*
*quaere fusius*

*Vide in proverbio*
*Perso sepiam*

Hic sepia non deest pedibus octo fulta, quae cum se peti animadvertit, atramentum, id est cruorem atrum loco rubri sanguinis effundit, os habet mollius, in argentariorum usum utile.

*20. L. Sepia*
*G. σηπία*
*H. xibia*
*I. Scuduil*

Huic sitis est loligo,[484] quae authore Plinio ex aquam sese efferendo volitat.

*21. L. loligo*
*G. τευθίς, τεῦθος*[485]
*H. Calamar*
*I. An luthurian*

Quibus polypus inconstantis hominis symbolum addatur.

*22 L. polypus*
*G. πολύπονς*
*H. Pulpa*

---

480.  'Ichthycolla' is crossed out in the R. margin and ἰχθυοκόλλα inserted from the L. margin.

481.  G. 'Maena' deleted in R. margin and μαίνα, μαίναις inserted from the L. margin.

482.  Scadán is herring. Pilser and serdin are pilchard.

483.  G. 'Egcheles' is crossed out in the R. margin and ἔλχελλος, ἔλχελις are inserted from the L. margin.

484.  Loligo crossed out and reinstated.

485.  G. Theutis crossed out and τευθίς and τεῦθος inserted.

The sturgeon has a glutinous hide and the same name is given   *Sturgeon*
also to the glue made from it.

The pilchard which, as far as we know, feeds on water and which   *Pilchard*
is commonly called maena[486] and scrombrus.[487] It abounds in
Ireland; but it is not a very fine fish, nor very well known.

*Pliny concerning the*   The types of the river fish with which the noble rivers and lakes   *Salmon*
*salmon, Book 9 Ch*    of Ireland abound should not be passed over. The salmon are
*18. The opinion of*    the best of the whole world, also the trout and the eels. See
*some people is that*   Huerta, p. 180, concerning the *carpion* trout also Pion.
*as a food it surpasses*
*all fish.*                                                                                                 *Trout*

                                                                                                            *Eel*
*Search more widely in*
*regard to the trout*

*See in the Persian*   Here, the cuttlefish is present, supported by eight legs. When it   *Cuttlefish*
*proverb the cuttle fish*   notices that it is being sought out it pours out a black liquid that
                       is black blood in place of red blood. It has a rather soft mouth
                       which is useful in the work of people handling silver.

With the cuttlefish is classed the squid[488] which, according to   *Squid*
Pliny,[489] flies by lifting itself from the water.

To these may be added the octopus,[490] a symbol of inconsistent   *Octopus*
man.

---

486.   Type of sea fish eaten by the poor. Pl 32.11.53.
487.   Type of tunny or mackerel.
488.   Loligo is also a type of cuttlefish.
489.   Pliny, 9, 45, 84.
490.   Pliny 9, 46, 85.

# C. XXXVIII
## Pisces plani

Quos hactenus retulimus pisces e genere longorum fere sunt.
De planis aliquot despiciamus.

| | |
|---|---|
| Raia in Ibernia tum nota tum frequens, cartilaginea, plana in tergo, caudaque spinas habet iis similes quas in rubo nasci videmus: unde graecis batos, idest rubus nominatur. | *23. L. Raia*<br>*G. βάτος*[491]<br>*H. Raya*<br>*I. Rathu* |
| Squatina quoque figura plana, cute usque adeo dura tegitur ut ea ligna et ebora poliantur. | *24. L. Squatina*<br>*G. ῥίνη*[492]<br>*H. lixa*<br>*I.* |
| Et utrisque raiae, et scatinae coitu scati raia, quae Graecis rhinobatas dicitur, oriri fertur. | *25. L. Squati raia*<br>*G. ῥινόβατος*[493]<br>*H.*<br>*I.* |
| Raiae minime dissimilis pastinaca, morsu hominbus noxia dicitur esse: in Iberniae littoribus aquis quandoque destituta praetereuntes conatur: sed plerumque diro omine suo, dum iactu, vel ictu collisa mordere, iacet. | *26. L. Pastinaca*<br>*G. τρυγών*[494]<br>*H.*<br>*I. gebeannach* |
| Passer Iberniae, qui et rhombus, psittaque vocatur, in optimo cibatu non iniuria habetur planus, et magnus. Rhombus et passer differunt. | *27. L. Passer*<br>*G. Psitta, rhombus,*<br>*ψῆττα, ῥόμβος*<br>*H. Rodaulio*<br>*I. Liothoig muiri*[495] |
| Eo multo minor, ei tamen nec figura, nec carnis sapore omnino dissimilis est solea Iberniae familiarissima. | *28. L. Solea*<br>*G.*<br>*H. lenguado*<br>*I. liothoig* |

491.　G. Batos deleted and βάτος inserted.
492.　G. Rhine deleted ῥίνη inserted.
493.　G. Rhinabato deleted and ῥινόβατος inserted.
494.　Trigon deleted.
495.　All crossed out in the L. margin.

# C. XXXVIII
## Flat Fish

Those described up to now are almost all of the genus long fish
Let us now take a look at flat fish.

The ray in Ireland is as well known as it is frequent, it is *Ray*
cartilaginous, flat, on its back and tail it has spines similar to
those we see arising from a blackberry bush from which it is
called by the Greeks *batos,* which is bramble.

The skate also has a flat shape and is covered by such a hard skin *Skate*
that wood and ivory are polished with it.

The skate-ray is held to spring from the mating of each type, the *Skate-ray*
skate and the ray. It is called *rhinobatos* by the Greeks.

The sting-ray is most like the ray; its bite is said to be poisonous *Sting-ray*
to men. Sometimes stranded in the coastal waters of Ireland it
tries to bite people passing by; but mostly this is a bad omen for
itself, as it lies crushed by a blow or a missile.

The Irish turbot, which is also called *rhombus* and *psitta,* is *Turbot*
rightly considered among the best food; it is flat and large. The
rhombus and the passer differ from one and other.

The sole is much smaller than the turbot, however neither in *Sole*
the shape nor in the taste of the flesh is it at all dissimilar; the
sole is very common in Ireland.

# C. XXXIX
## Aquatilia crustis intecta

*29. L. Cancer*
*G. καρκίνος*[496]
*H. Cangrejo*
*I. Partain*

Aquatilia non nulla crustis roborata contemplemur. Eorum notissimus est Ibernis cancer fragili testa vestitus, cuius species aliquot a Plinio referuntur, ut carabus, astacus, maia, pagarus, heracleoticus, leo.

*L. Lucusta*
*G. κάραβος*[497]
*H. Langosta*
*I. gliamach*

Haud equidem ignoro, aliquos in ea esse sententia ut vellint carabum astacum et squillam sive scillam esse nostram locustam in Ibernia frequentissimam non satis firma crusta defensam, communi cancro longiorem praeter quam et aliae sunt eadem figura minimae.

*30. L. Hericius*
*G. ἐχῖνος*[498]
*H. Erizo*
*I Cuan mairi*

Insulae nostrae echini sive hericii (sunt et alarium regionum) crusta, spinis defenduntur (unde propter similitudinem *ἐχινομήτραι* acastaneae tegumine nomen acceperunt, orificio scopulis haerent, illis progredi est in orbem volvi),[499] eorum illi quibus spinae longissimae, calices minimi sunt echinometrae appellantur.

# C. XL
## Uniones, et Conchae

*31. L. Testudo*
*G. χελώνη*[500]
*H. Galapago*
*I.*

Multo firmiore munimine, hoc est silicum duritia propugnantur alia aquatilia, ut testudo marinum animal, quadrupes durissima concha instar scuti dorsum tectum in Ibernia perraro visum.

---

496.   G. Carabus deleted and καρκίνος inserted.
497.   G. Carcinos deleted and κάραβος inserted.
498.   G. Echinos deleted and ἐχῖνος inserted next to it.
499.   The section in brackets is inserted from the R. margin. Crossed out in the centre of this insertion, is the following 'ita lenis procedi est in orbem volvi alteri?'.
500.   G. Chelone deleted and χελώνη inserted.

# C. XXXIX
## Sea creatures protected by shells[501]

*Crab*

Let us consider some sea creatures strengthened by shells. The most well known to Irish people is the crab, clothed in a fragile shell, some species of which are referred to by Pliny such as the carabus (common crab), astacus (crayfish), maia (spider crab), pagurus (hermit crab), heracleoticus (Heraclean crab), leo (lion crab).[502]

*Lobster*

I am not unaware that there are those who would have it that the crab, crayfish, and the prawn or shrimp are our lobster in Ireland. Lobsters are very common, protected by a rather weak shell; they are longer than the crab; besides these, there are others, very small, which have the same shape.

*Sea-urchin*

The sea urchins or hericii of our island, like those of other places, are protected by a shell and spines.[503] (From this, on account of the similarity, they have received their name from the covering of the chestnut. They stick to rocks by an opening and they move forward by turning in a circle.) Those of them with the longest spines and smallest cups are called echinometrae.[504]

*Large sea-urchin*

# C. XL
## Pearls and shellfish

*Turtle*

Other sea animals are defended by a much stronger protection, which has the hardness of flint; such is the turtle, a quadruped with its back protected by a very hard shell, as with a shield. It is very rarely seen in Ireland.

---

501. Silicu duritia induta is crossed out.
502. Pliny 9, 51, 97–99.
503. The section in brackets is inserted from the R. margin. Crossed out in the centre of this insertion is the following, 'ita lenis proredi est in orbem volvi alteri'.
504. Echinometrae are the largest of the sea urchins.

*32. L. Margárita*
*G. μαργαρίτης*[505]
*H. Nacar*
*I. Maither Pearla*

Atque nobilissimae concharum margaritae: quae in Iberniae littoribus binae inveniuntur conchae, longitudine, ut plurimum palmi, latitudine trium circiter digitorum, externa facie nigrae, interna candidae, mollem carnem (sicut ostreae) continentes. In hac carne nuclei quidam idest lapilli duri inveniuntur:

*L. Unio*
*G. μάργαρον*[506]
*H. Aljofar*
*I. Piarla*

quorum minores eodem nomine margaritae, maiores in summo pretio semper habiti uniones nominantur. Horum aliqui e concha exeunt multiplici cute induti, quae a peritis purgantur, et exuuntur, ut non improprie calla concharum dicantur: alii rubro, alii subnigro, alii candido colore sunt. Eorum dos omnis in candore, magnitudine laevore pondere, rebus haud promptis constitit.

[507]

Hinc Iberniae pectines magnitudine praestant, striati (sic) esu iucundi.

*33. L. Pectines*
*G. κάυη κάυνη*[508]
*H.*
*I. Macmuirin*

Aliae similimae, sed minores, et altera aure maiore in Ibernia quoque frequentes, aures marinae vocantur.

*34. L. Patell*
*auris marina*
*H.*
*I. Cluosin*

Huc accedunt tellinae etiam concha striatae, et in cibum utiles, sed minores.

*35. L. Tellina*
*G. τελλίνη*[509]
*H. Ruocan grubán,*
*ruacan*

Silentio praetereundi non sunt mytili Ibernis satis noti nigra, et oblongo concha palato non ingrati, alteri maiores, alteri minores.

*36. L. Mytilus*
*G. μύαξ μυτίλος*
*H. Almeja*
*I. eascán, musla*

---

505. G. Margarites deleted in L. margin and μαργαρίτης inserted from R. margin. μαργαρίτης deleted in R. margin.
506. Margarito crossed out, μάργαρον inserted, margarites deleted.
507. Crossed out in L. margin, faochog traich umbilicus, corcaer purpura acutissim ae choncae. I. Cian mhairi vide ungues camcoinil vide veenia, tamsalt mhairc, dubh liasc feamnach.
508. Lhiatucae crossed out? G. Chame and Cheme deleted, κάυη, κάυνη inserted.
509. G. Tellinae deleted and τελλίνη inserted.

*Mother of Pearl* And the most noble of the shellfish are the mother of pearl, which are found as bivalves on the shores of Ireland, in length at most a palm's breath, in width, about three fingers; the outside surface is black, the inside, shining white, containing soft flesh (just as that of an oyster). In this flesh are found certain nuclei, i.e. hard little stones of which the smaller have the same name, *Pearl* pearl; the larger, always considered the most precious, are called uniones (single large pearls). Some of these come out from the shell covered in a multi-layered skin which is washed off by skilled people and stripped so that, not unsuitably, they are called the hard skins of the shells: some are red, some blackish, some shining white in colour. Their whole quality consists in their whiteness, size, smoothness, and weight, things that are not obvious.

Thus, the scallops of Ireland are outstanding in size, they are *Scallop* ridged and pleasant to eat.

Others are similar but smaller, and another larger than an ear *Sea ear* are frequent in Ireland; they are called sea ears.

Next come the cockles also with striated shell and useful to eat *Cockle* but smaller.

The mussels must not be passed over in silence, well known *Mussel* in Ireland, black with an oblong shell, not unpleasant on the palate, some are larger, some are smaller.

Cochlae littoribus Iberniae familiarissimae[510] tum domicilium
suum dorso portant binis corniculis iter praeteutantes, quod
oculis destituuntur, tum in id se contrahunt totas:

*37. L. Cochlea*
*G. κοχλίας*[511]
*H. Caradcol*
*I. Piachain*

Lepades[512] saxatiles rupibus carnea parte firmiter haerent,
altera parte firma concha muniuntur. Arist. hist. lib. 4 c.4.
*Sunt item, quibus altera pars superficiei detecta carnem ostendat,*
*ut* patellae. Rusus.
*Patella etiam saxis absolvi, in pastumque ferrisolita est.*

*38. L. Cochlea*
*saxatilis, patella*
*G. λεπάς*
*H.*
*I. Baerneach*

Marinae glandes, minutulae conchae, sed plurimum candidae
adiungantur: Haec concharum genera meminisse, sufficit:
omnes, quas Ibernia generat, recensere, taediosum fuisset. Tot
enim ibi visuntur, ut in eis mira naturae varietas conspiciatur. Illis
tot colorum differentiae, tot corporum figurae, planae, longae,
concavae, in dorsum elatae, lunatae, in orbem circumactae,
dimidio orbe caesae, laeves, rugatae, denticulatae, striatae.

*39. L. glans*
*G. Βάλανος*[513]
*H.*
*I. Moedeoigh*
*Bollán an tuberculum*

## C. XLI
### Spongia, et quibus aquatilibus caret Ibernia

*40. L. Spongia*
*G. σπογγιά*[514]
*Σπόγγος*
*H. Asponja*
*I. Spunc*

Hanc aquatilium considerationem ut absolvamus, naturam
spongiae marinae, quam Ibernia plurimam gignit, scrutemur.
Ea nec animal, nec frutex, sed inter utrumque naturam tertiam
habens cum aliis tum Plinio indicatur. Quem rerum ordinem
ξωόφυτον[515] id est vivam plantam Graeci vocant.

---

510. After 'familiarissimae', 'altera' is crossed out.
511. G. Cochlias deleted in R. margin and κοχλίας inserted from the L. margin.
512. 'Altera' deleted and 'lepades' inesrted above the line in L. margin. In L. margin λεπάς, G., L. Patella, H. I. Baerneach are crossed out.
513. G. Balanae deleted and Βάλανος inserted.
514. 'Spoggos' is deleted in L. margin and σπολλιά Σπόγγος inserted from R. margin. G. ωόψυτυν is crossed out above these two.
515. Zoophyton is deleted in the text and ωόψυτυν inserted from the L. margin.

Periwinkles are very familiar on the shores of Ireland. *Periwinkle*
Sometimes, they each carry their dwelling on their back, feeling
their way as they go with two little horns because they have no
eyes; on other occasions they withdraw themselves completely
back into the shell.

Limpets stick firmly to rocks by the fleshy part; on the other part *Limpet*
they are protected by a strong shell. Arist. *hist. lib.* 4 c.4: '*They
are also creatures of which one side of its surface being uncovered
displays flesh, such as limpets.*' Again: '*The limpet, when it is prised
from the rocks, is usually turned into food.*'

*ea acorn*  Sea acorns are tiny shellfish, but for the most part white shellfish
may be added. It is enough to mention these types of shellfish. To
recount all the shellfish that Ireland produces would be tiresome.
For so many are seen there, that in them the wonderful variety of
*bollán the*  nature is seen. There are so many different types among them of
*berculum?*  colour, so many shapes of body: flat, long, concave, lifted up on
the back, lunate, fashioned round into an orb, cut into half an orb,
smooth, wrinkled, denticulate, striated.

# C. XLI
## Sponges, and the marine forms
## which are lacking in Ireland

*ponge*  In order that we may finish this consideration of aquatic
creatures let us examine the nature of the marine sponge which
grows a lot in Ireland. That it is neither animal or shrub but has
a third nature between the two is indicated by Pliny[516] as well
as others. That order of things the Greeks call Ζωόφυτον, that
is, a living plant. The sponge is born on marine rocks, is seen to
have strong sensation because it pulls itself back when touched
by the person plucking it so that with much greater difficulty
it is pulled from the rocks. We use it for cleaning clothes and
tables as Martial[517] records:

---

516.   Pliny, 9,69,148.
517.   Martial 14, 144.

Spongia in saxis marinis nascitur, sensu pollere videtur, quia ab avulsore tacta ita se contrahit, ut multo difficilius e lapide abstrahatur. Ea ad tergendas vestes, atque mensas utimur, ut meminit Martialis.

*Haec tibi forte datur tergendis spongia mensis*
*Utilis expresso cum levis imbre tumet*

Spongiarum tria genera Plinio numeratur: spissum, praedurum et asperum nomine tragos; Mollius, et minus spissum manon; tenue; densumque acilleum.

His aquatilium Iberniae reputationi summam manum imponamus: nam nostrum institutum fuit, ostendere, multo plura esse iis quae Gyraldus nominavit; non vero omnia perstringere, quod foret longissimum, quia illius insulae oceano, et aqua dulci, utique non omnia, pleraque tamen alibi cognita (modo crocodilos, et alias nocuas belluas demas) gignuntur.

## C. XLII[518]
### Viventia quae sensu carent

*Dic de labrusca id est vite Sylvestri ex Plin. l. 14 c. 16*

Hucusque naturas animantium, quae in Ibernia locali motu agitantur, sentiendique vi pollent, enarravimus. Alia viventia, quae animam quidem; non tamen moventem loco, vel sensitivam; at nutrientem habent, super sunt, ut arbores, frutices herbae. Quibus est ea insula copiosissima. Quamobrem singulorum vel nomina comprehendere, longissimum, et plusquam unius scriptoris opus fuisset . . . Ne vero a tam uberi et amoeno memorabilium rerum horto ieiuni egrediamur nonnihil dicere putavi ab arboribus frugiferis auspicans.

---

518.    XXXX is crossed out.

*This sponge by chance is given to you for the purpose of cleaning tables. It is useful, when it is puffed up light after the water has been squeezed out.*

Pliny[519] enumerates three types of sponge: the dense type, very hard and rough, a goat-thorn by name: a softer and less dense, the loose sponge, and a fine and dense sponge, the achilleus.[520]

With these let us put the final touch to our consideration of the aquatic creatures of Ireland: it was my intention to show that they are much more numerous than those which Gyraldus[521] mentioned; I decided not to touch upon all indeed, because it would be too long, because in the ocean of that island, and in the fresh water, while not all, most creatures that are known elsewhere (leaving out crocodiles and other harmful beasts) are born here.

## C. XLII
### Living creatures which lack feeling

*ell about the*
*abrusca or wild vine*
*Pl. book 14,16.*

Thus far we have discussed the natures of creatures in Ireland which move from place to place by motion, and who have the power of perception. Other living creatures remain that have a lifeforce which has not however the ability of movement or feeling but which does have the ability to feed, such as trees, plants and grasses, with which this island is copiously endowed. For this reason to include the names of every single one would be too long and too much for the work of a single writer. Lest, truly, we should go away hungry from so fruitful and pleasant a garden of memorable things, I decided to say something, beginning with the fruit-bearing trees.

---

519.　Pliny 9, 69, 148.
520.　Achilleos a medicinal plant said to have been discovered by Achilles Pl. 25.5.19
521.　Philip O'Sullivan is pointing out the much greater detail of his account of the natural history of Ireland than that of Gyraldus, thus implying that Philip's account is more reliable and so his criticism of Gyraldus's history deserves respect.

# C. XLIII
## Arbores frugiferae

Principio in Ibernia vites, vinaque gigni, prisci scriptores memoriae prodiderunt: nostra tamen tempestate illae perquam paucae creantur: harum[523] vero uberrimus est proventus, sed peregre, ut ex Hispanis Gallia, et insulis fortunatis advectarum. *Dic de olivia quae est in Ibernia seri copeta. Vide orchas in an: etc.*

*1. L. Vitis*
*G. ἄμπελος*[522]
*H. vid*
*I. Finuir*

Malus autem, sive pomus hic varia malorum genera producit, alia dulcia, alia acria.

*2. L. Malus pomus*
*G. μηλίς*[524]
*H. Manzana*
*Ib. Abhall*

*Dic de ficu quae est*
*sata Pontanae in*
*horto*
*Mespilus ton osculithi*
*vide annot.*

Pyri quoque pyra proferunt inde dicta quod ad similitudinem flammae elato in acumen tendant. Utrorumque generi mespila sorbaque cum aliis adnumerantur.

*3. L. Pyrus*
*G. ἄπιος*[525]
*H. Peral*
*I. Peri*

*4. L. Ficus*
*G. συκή*
*H. Higuera*
*I. Figi*

Ficus non est praetereunda, quamvis non frequentissima sed fructu[526] grata.

*5. L. Prunus*
*G. κοκκύμηλος*[527]
*H. Liruelo*
*I. Droin*

At frequens spinus arbor sylvestria pruna generat, prunorum alia acria alia dulcia.
*Vide prunus et spinus apud calep vide de muchoir (plum tree) paluirus ex plin1. 13 c.7 sc 18. Dic de donnuir.*

---

522. G. Ampelos deleted and ἄμπελος inserted above it.
523. 'Huiusque' crossed out and 'harum' inserted above the line.
524. G. an illegible word is deleted in R. margin and μηλίς inserted from L.margin.
525. G. Opios is crossed out in R. margin and ἄπιος inserted above it.
526. Crossed out in R. margin 'paliurus ut puto skeach, fert paliurusque folia?' Plin1.24 c.13, dic de drean, droin et eius fructu.
527. Coccoimelos deleted in L. margin and κοκκύληλο inserted from the R. margin.

# C. XLIII
## Fruit-bearing trees

In the beginning, early writers recorded that the vine and wines *Vine* were produced in Ireland: however, there are very few of them produced in our time. Truly the product of the vine is there in abundance, imported from abroad such as from Spain, France and from the Canary Islands. *Mention the olive which has begun to be planted in Ireland. See Orchas (a type of olive)*

The apple tree, or the pomus, produces different types of apple *Apple* here, some sweet, some sour.
*Tell about the olive which has begun to be cultivated in Ireland. See Orchas (type of Olive where noted) P101*

Pear trees also bring forth pears, so called because like a flame *Pear* it proceeds from a broad base to a point. The medlar and the sorb apple are enumerated in the class of both of these along with others.
*Tell about the fig which was planted in a garden in Drogheda.*

*g*

The fig tree must not be passed over: although not the most frequent, the fruit is pleasant.

*lackthorn*

But the blackthorn, a frequent tree of the woods, grows the sloe: some of the sloes are bitter, some are sweet.
*See Prunus and spinus in Calepinus. See about the plum tree. Mention muchoir (deleted), Christ thorn from Pliny 1.13 c7 sc. 18, mention Donnuir*

*6. L. Spinus*[528]  Spina vero sylvestris Graecis acantha aculeis acutis surgens
*H. Spino*          appendix vocatur ea baccas puniceo colore profert vocatas
*G. ἄκανθα*         appendices vocatur eam baccas puniceo colore profert vocatas
*I. (Sceach deleted)* appendices.
                   Plin 1.24 c.12 13 & 14 ibi Dalecamp. cc.1.16.37. skeachor
                   Hanma (kieca.).

*7. L. Cerasus*[529]  Cerasus cerasa generat, colore rubra acetosa dulcedine. Plin. 1
*G. κεραοός*          16.25.
*H. Zeraza*
*I. selin*

530

*8. L. Cornus*        (Corna cornus generat, corno masculae medullanulla. Plin. 1
*G. kρavía*[531]      16.25[532]
*H. Cerezo sylverstre*
*I. Carrhinn*
*(caorthann)*[533]

*9. L. Lentiscus*     Quaedam lentiscus arbuscola vel frutex acinos producit,
*G. Xivos*[534]       dum maturescunt, ruffos vei maturescuerint nigros eiusque
*H. Lentisco*         iucundissimos.
*I. Freachain*

*10. L. Arbutus*      Arbutus fructum dictum Graecis praebet est[535] rubercolore,
*G. κόμαρος*[536]     mollisque
*H.*
*I. subhcroabh*

---

528.  I. Sceach deleted.
529.  Cerasos deleted.
530.  Crossed out in this paragraph: 'tota vera spinus' is crossed out and
      'spina vero sylvestris' above the line. 'Alios fructurs procreatsatis nota'
      is crossed out and 'appendix vocatur' inserted above it.
531.  Crania deleted and kravía inserted.
532.  Crossed out between Cornus and Arbutus with a line drawn around
      it: 'Quadum lentiscus arbuscuila vel frutex acinos producit dum
      maturescunt, ruffos ubi maturuerint nigros esuque iucundissimos.'
533.  Carhin crossed out.
534.  G. Chinos deleted and Xivos inserted.
535.  In the paragraph concerning arbutus there is a lot of illegible crossing
      out between 'praebet' and 'est'.
536.  Comerus deleted and κόμαρος inserted.

*1orn*

A wild thorn tree, called achanta by the Greeks, growing with sharp thorns, is called the barberry bush. It produces berries purple in colour, called barberry (Plin. book.24 chapters.12, 13, 14 ibi. Dalecampius. chapters.1.16.17).

*1erry*

The cherry tree provides cherries red in colour and pleasant with a sharp sweetness.

*owan*

The rowan begets cornel berries. There is no pith in the centre of the cornel cherry. Pliny, book 16, chapter 25.

*1astic*

The mastic tree, a little tree or shrub, produces berries. While they are ripening the berries are reddish but when ripe, are black and most pleasant to eat.

*rbutus*[537]

The wild arbutus produces a fruit called by the Greeks μεμυλον.[538] It is red in colour and soft.

---

537.   Subhcraobh (sú craobh) is a raspberry in Irish and is different from the arbutus tree which produces strawberry-like berries.
538.   μεμυλον was crossed out in manuscript.

*11. L. Fragum*[539]
*H. Fresas Terrestes*
*Mielgado, mayeuta*
*I. Sumbh talbhuin*

Fraga[540] in herbis nascuntur,[541] unde frutus terrae nominantur.

*12. L. Rubus*
*G. βάτος*
*H. Zarza*
*I. Dris*

His maiora mora in rubo, frutice aspero, notissimoque crescant: quibus trini sunt colores: primus candidus, inde rubens, maturis niger.

*13. L. Rhamnus*
*G. ράμνος*
*H. Cambron*
*I. spin,*[543] *spinog eius*
*fructurs vide annot*

Ex ruborum[542] tamen genere rhamnus.

*26. L. Tilia*[544]
*G. φιλύρα*[546]
*H. Teja*
*I. Lambhinn*

Tilia[545] extima cute fragili, interioribus tunicis, quae inter extremam corticem, et lignum mediae sunt flexibilibus. Vide Plin. & calep.

Salix hic[547] a saliendo (sicuti volunt) id est crescendi celeritate dicta in amnium ripis frequens.

*27. L. Salix*
*G. ἰτέα*
*H. Sauze*
*I. saileach (saileach)*

Aquoris quoque locis alnus gaudet quod enim amne alatur, nomen hebere videtur.

*28. L. Alnus*
*G. κλήθρα*[548]
*H.*
*I. Fearnoig*

---

539.   G. Coma deleted.
540.   'Unedonibus similis' is crossed out before 'fraga'.
541.   'Humi nascentibus' is crossed out after 'nascuntur'.
542.   Hoc maior et fructuosior crossed out before 'exruborum', and 'ex albo flore' crossed out after 'rhamnus'.
543.   Three illegible words deleted spin et spinan also crossed out.
544.   There is no explanation for the jump in numbers.
545.   This paragraph is inserted from the L. margin. After 'tilia' modica proceritate beithi are legible and four words illegible.
546.   G. Philyra deleted and φιλύρα inserted.
547.   After 'hic' plurima is deleted.
548.   Elebore deleted.

*Strawberry*

Strawberries grow in grasses from which the fruit is called fruit of the earth.

*Blackberry*

Greater than these is the blackberry, it grows on the bramble bush on a rough bush and well known. The berries have three colours: first white, then red, and black when ripe.

*Blackthorn*

However, the blackthorn is of the bramble family.

*Lime*

The lime, with the outermost bark fragile, has a flexible interior membrane between the outside cortex and the wood of the centre. See Plin. & Calep.

The willow is so called from springing (just as they are fond *Willow* of doing); that is, from the swiftness of its growth. It is usually found on the banks of rivers.

The alder also loves wet places. It seems to be named by the fact *Alder* that it is nourished by the river.

Arundo intus corpore cavo, superne tenui ligno, geniculata
frequentibus nodis distincta, coma panniculae crassiore inter
aquaticos calamos principatum sibi vendicat.[549]
*Cremacordium glearan an?*

*29. L. Arundo*
*G. κάλαμος*
*H. Cana*
*I. Bearrach*

*Betulla Plin. 1. 16c. 18; vide beithi.*
Suber Hisp. Alcornoque. crabhan coirc Plin. 1. 16. c. 8.

*Limon & naranja nascuntur in Iberna vide citreum apud Plin.*
De edera einean, et Helice tahelinn vide Plin 1.16.c.34 et de
smilace c.35. Crabhan oangna. Plin. L.16. c.17 idem c. 24 folia
ramulosa ulmo.

*Cuilinn maol quaere*
*nomen*

Cuilinn crisp. Avpix as. arbor semper virens aqui-folium
φελλόδρος arbor semper virens vide Dalicamp in c.6.L.16 Plin.
Idem Plin. L. 15 c.24 bacca aquifolio sine succo, folia pungentia
sunt aquifoliis. Plinio etiam aquifolia vocatur, et Dalecampis ex
ea fit viscus.

Feothanan vide eringe et eringion Plin. L.22.c.7 vide et capa
sequentia.

Siler calepino pumila salix. vide Virg. 2. Georg.

Arnus sylvestris fraxinus calep. vide Virg. Orchas in accituno,
orcal, o judiega, nebri; vide ochites in dicto radius apud Calep.
radii obivae oblongae. Calep. pausia genus olivae Virg. & Calep.
volema, pyra grandia. Calep. et Virg.

---

549.  After 'vindicat' the following two paragraphs are deleted. Hedera
      cum se sustinere non possit aliis arboribus atque rersus innixa surgit.
      Malum punicum vel grand um est in Ibernia in qua vocatur ubhuall
      grain vide Plin.1.13 c.19. L. Hederea, G. Citros, cistos, H. Yerda, I.
      (illegible word delete), faith helinn.

The reed has a hollow inside and on top a light wood and it is knotty with frequent nodes, with plume more coarse-grained than rags. Among the aquatic reeds it claims to be the most outstanding. *Cremacordium is glearan?*[550]

Betulla (birch) see Plin. L.16 c.18, see beithi (modern Ir. beith or beithe) birch. Suber (Cork Oak) in Spanish Alcornoque. Crabhan Coirc (Cork tree) Pl. L.16 c.8 §34 §58.

*Lemon and Orange grow in Ireland. See Citrus Fruit in Pliny.*
Concerning Edera Einean (Modern Ir. eidhneán); ivy helix; a type of ivy. Tahelin (?Táith-fhéithleall) honeysuckle, see Pliny book.16 c.34. also concerning bindweed, Pliny book.16 c.35. Crabhann oangna, Pliny. book 16 c.17. Also Pliny book 16 c.24. Branching leaves on the elm, Pliny. 16.76.

*Cuillin maol check the ame* Cuilinn crisp. Avipix, an evergreen tree, sharp-pointed leaves φελλόδρος, an evergreen tree. See Delacampus in c.6. book 16 Plin. The same Pliny 15.c.24. The holm oak has a berry without sap. The holm oak has prickly leaves. It is also called a holm oak by Pliny and Delacampus. Bird lime is made from it.

Feothan see eringe or eringion (thistle), see Pl. 22,7 and the following chapters.

Siler (dwarf osier), pumila salix (dwarf osiers) in Calepinus. See Virgil *Georg.* 2,111.

The mountain ash is the wild ash in Calepinus, see Virg. 2 *Georg.* 110.
The orchas (a round olive) *aceituna* Spanish for olive, orcal. Also called Jewish Nebris. See Ochites radius [in the dictionary of] Calepinus. Radii are oblong olives in Calepinus. Pausia a type of olive *Georg.* 2, 84–88. Volem, a large pear.

---

550.　O'Sullivan here seems to use a technique used by Pliny in Book I of his Natural History. Pliny just writes down lists of objects which seem to be random at times but explains them in detail in a later book or chapter. O'Sullivan possibly is just making notes here and in the next two pages.

Vitex, Hisp. Sauzgatillo. Ib. broin leog shaili.
Vide Calepinum. oret saileog.

Morus, Hisp. moral Iber. Crabhonn smeara et morus est in Ibernia.

Oliva in Ibernia est coepta seri in Tomonia.[551]

Iber. gloiriem Hisp. lirio cardeno, iris et hyacinthus

Nebris. apud quem vide plura

Aitin[552] vide Erica[553] froach vide Brya, mirice, tamarix.

Genista Hisp. retama. Ib. gealceach & Hisp. Hinestra
Apiastrum Hisp. torongil an crebhel?

Socuilii folium, meacabrii radix.

Bainni liana vide Githymalus

Lotos arbor est baccifera Plin. L. 15. c.24

Sceabh croinn

Aitin francach

Inis faithlinn insula in leno lacu

Membrillo[554] Iber. quince ex Anglo sermone.
Haec arbor sata tulit fructum in Connachta

in horta Joannis Burki sata fuit a medico[555] cluthuran tuber terrae.

Taxus. Ib. ur Plin. 1.16 c.20 in fine. Diligentissime videndus Dalecampius in c.13 eiusdem libri amigdalus est in Ibernia sativa.

---

551.   'Brean train' deleted after tomonia.
552.   Atan crossed out before aitin.
553.   et eryxysception deleted after Erica.
554.   After membrillo, 'vide nebr'. Followed by five illegible words.
555.   After medico vide? catoneum is crossed out.

Vitex (chaste tree) in Spanish Sauzgatillo, in Irish Broin leog shail; see Calepinus, sally.

Morus: in Spanish, moral; in Irish, crann smeara. The mulberry is also in Ireland.

The olive has begun to be sown in Ireland, in Thomond.

In Irish the gloriem (glóirian) blue iris; in Spanish lirio cardena, the iris and hyacinth.

Nebris seek more in his writings.

Aitinn (aiteann) furze, see erica (broom), fraoch (heather).

See Broom mirice, tamarix (tamarisc) Pl. 13,21,37.

Genista in Spanish retama; Irish *gealcach* and in Spanish hinestra. Apiastrum in Spanish, torongil or crebhel? (?broom)

*Sorcuilli* folium *macabui* radix (carrot).

Bainni liana, see Githymaius.

Lotus is a berry-bearing tree. Pliny, 15.24.

Sceabh croinn.

*Aitin franncach* french furze.

Inis Faithlinn, an island in Loch Lein.

Membrillo in Spanish, quince in English. This tree was planted and bore fruit in Connacht.

In the garden of John Burke, the cluthuran, an earth nut was planted by a doctor.

Taxus (a yew tree) in Irish ur (iubhar) Pl. 16,20 at the end. Dalecampius in chapter 13 of the same book, should be looked up diligently.
The almond tree is planted in Ireland.

Coilin moal. quae re eius nomen

Tamarix est in Ibernia sativa saltem.
*Dic de moro arbore et eius frcutu nam nascitur in Ibernia dic de muchoir vel muchor*

Profert folliculos[557] rectis aculeis ramos spargens. Ex eo spineam coronam quae salutis humanae Christi sacrum caput crudelissime vulneravit, textam fuisse, sunt authores. Eundem in incuborum nocturnorum, lemorum, strigum, veneficia damnaque efficax amuletum esse credidisse et ob id in parentationibus ante flores collocare, soliti fuisse, ethnici feruntur. Quamobrem spinam ianalem esse nonnulli volunt Ovidii carmine celebratam:
*Sic fatus, spinam, quae tristis pellere posset*
*A foribus noxas (haec erat alba) redit*

Glandem quoque durum et robur fert. Ea prisci mortales vitam tolerabant antequam luxus genus humanum corrumperet, tot obsoniorum genera ventri excogitans: quo primum[560] illud saeculum nova aetate foelicius fuerit, attestante divo Boetio
*Foelix nimium prior aetas contenta fidelibus arvis:*
*Nec inerti perdita luxu facili quae sera solebat ieiunia solvere glande*

*14. L. Robur[559]*
*H. Roble*
*I. Daer[561]*

Castaneae: seu glandium, seu nucum[562] ex geniri adnumerentur, echinato calice vestitae hic minime negantur. Meas mairineach castaneae sunt in Ultonica sylva apud Claun eth Bui.

*15. L. Castanea*
*G. κάστavον[563]*
*H. Castano Castana*
*I. Castana*

---

556. Crossed out in margin nine illegible words, finishes with vide Plin.
557. 'Profert folliculos' inserted above the line. Crossed out: 'E flore suas etiam baccas esu non ingratas suppeditat primum virides inde rubras denique subnigras'.
558. Crossed out in L. margin: 'Dic de fructu rubro quod profert muchoir. G. fructus rubens habet grana dic et de corróg'.
559. Lomán deleted.
560. 'Quo primum' to 'solvere glande' is inserted from the L. margin.
561. Darrach deleted – illegible words deleted after Robla.
562. Ex is inserted above the line.
563. G. Castana deleted and κάστavον inserted.

Bald holly, check its name.

The tamarisk at all events is sown in Ireland. *Tell about the morus (mulberry tree) and its fruit for it grows in Ireland. Tell also of muchoir or mucor (mulberry) in Spanish escaramojo.* Spreading its branches, it brings forth follicles with straight points. It is recorded that from that plant the crown of thorns was made which wounded in a most cruel way the sacred head of Christ, the salvation of the human race. Pagans are held to have believed the same to be an efficient charm against poison and loss caused by nocturnal incubi and witches. And because of this they are accustomed to put tamarisk in front of flowers at funeral obsequies. Therefore, some wish the thorn of Janus to be celebrated by the poem of Ovid.[564]
*Having spoken, he gave her the thorn (it was white), which could drive away harmful things from the door.*

The oak also bears a hard acorn. With it mortals of old used to support life before luxury corrupted the human race, thinking up so many types of food for the stomach: in this way that first age was happier than the modern age, as is attested to by Holy Boethius[565] *The previous age, overly happy, contented with the trusted fields nor ruined by idle luxury, was wont to allay its late fasting with the acorn, easy to gather.* *Oak*

The chestnuts, from a type of horse chestnut, must be counted among the family of either acorns or nuts; they are covered with a prickly cup and not denied here. The meas mairineach (meas marthanach) are chestnuts found in the Ulster forest among the Claun eth Bui (clannboy) *Chestnut*

---

564.   Ovid *Fasti*, Book 6, 129–30.
565.   Boethius, *De Consolatione Philosophiae*, Book 2, Metrum 5, Line 1–2.

Pinus non deest, maxime in Conkeinia sylva arbor satis alta: in qua nux ominum maxima pinea crescit altissime suspensa intus exiles quidem nucleos, sed plurimos lacunatis toris includens, singulis durissimis tunicis vestitos. Ex ea arbore resina humor tenax, et in multos casus utilis fluit et hodie crescit in Ibernia in aliquibus locis frequens.

*16. L. Pinus*
*G. πίτυς*[566]
*H. Pino*
*I Crann rosin*[567]
*Vide gaeda et picea*

Nucum autem optimam iuglandem nux arbor praebet, geminis operimentis protectam foris calyce intus ligneo putamine.

*17. L. nux iuglans*
*G. κάρυον*[568]
*H. Nunez*
*Ib cno francach*

Corylus avellanas nuces copiosissime reddit.

*18. L. Corylus*[569]
*G. κόρυλος*
*H. Avellano*
*I. Coll Dic dei*[571]

## C. XLIV[570]
### Aliae arbores

*Dic de amygdalequae est in Ibernia sativa*

His alias arbores adiungere possumus quae quamvis fructus in humanum alimentum idoneos non ferant, in usus tamen plurimos sunt perutiles.

*19. L. Fraxinus*
*G. μελία*
*H. Frezno*
*Ib Foin Seoig*

Primum quidem fraxinus arbor procera teres est materies ad multa quam aptissima.

*20. L. Acer*
*G. σφένδαμος*
*H. Azre*
*I. foin seail*

Accedit acer operum elegantia, subtilitate que commendabile. Vide zygias et carpinus apud Plin. 1.16. c. 15. Huius triagenera quae sunt in Ibernia et alia.

*Buxus omnino agrestis vez oleastrum brean carabhan Pl L.16 c.16.*[572]

---

566.    G. Pitys deleted and πίτυς inserted.
567.    Gimhos crossed out and giumhius inserted.
568.    G. Carryon deleted in R. margin and κάρυον inserted from the L. margin.
569.    Corylavel deleted after Corylus.
570.    XXXX deleted before XLIV.
571.    I. Coll. deleted and reinstated.
572.    This line is inserted above the line before 'ne buxum'.

There is no lack of pine, quite a high tree, mostly in the *Pine* Conkeinia forest, on which the biggest pine cone of all grows, suspended very high up, enclosing in hollow nodes, kernels which are slender but numerous, separately encased in very hard coverings. From that tree there flows resin, a sticky liquid useful for many jobs. Even today it grows densely in Ireland in some places. *See gaeda (pinewood) and picea (spruce or fir).*

However the walnut tree produces the walnut, the best of nuts, *Walnut* protected with two coverings, a cup outside and a woody shell inside.

The hazel produces hazelnuts in great abundance.[573]          *Hazel*

## C. XLIV
## Other Trees

*Mention the almond* With these, we are able to join other trees which, although *tree which is planted* they do not bear fruits suitable for human consumption, are *in Ireland* however useful for many jobs.

*Ash* First indeed is the ash tree, tall and smooth, it is a timber most suitable for many purposes.

*Maple* The maple comes next, commendable for the elegance and fineness of its material. See Hornbeam and Carpinus, according to Pliny 16, c. 26, section 66. Of this there are three kinds in Ireland and elsewhere.

---

573.    Abella (now Avella Vecchia), a town in Italy, was very productive of nuts. Vir *Aen.* 7, 740.

| | |
|---|---|
| *21. L. Buxus*<br>*G. πύξος*<br>*H. Box*<br>*I. Busca*[575] | Ne buxum praetereamus, adeo ponderosam ut in aqua subsidat, duram ut cariem non sentiat. Ex ea in Ibernia lapilli precatorii fiebant.[574] |
| *22. L. Populus*<br>*G. λεύκη*[576]<br>*H. Alamo*<br>*Ib Crithir* | Populus alba, et nigra, et lybica, sequatur, vide Plin. 1.16 c.23 idem c.24. Populi folia pediculo tremulo, et solius inter se crepitantia, Eius tria genera.<br>*Examina. Populus alba est certe crithir et populus nigra.* |
| *23. L. Laurus*<br>*G. δάφνη*[577]<br>*H. Laurel*<br>*Ib Lauruis* | Laurus, quae perpetua coma viret, suas baccas mittens adnumeratur. |
| *L. Sambucus*<br>*G. ἀκτῆ*[578]<br>*H. Sauco*<br>*Ib trium et troman* | Sambucus foliis amplis, flore odorifero baccis, dum maturescunt, rubris, cum maturuerint, nigris, medulla alba, et spongiosa, cortice plumbei coloris crasso multis nodis distinguitur.            [579] |

580

Paliurus baccas habet, foliaque spinosa, ut meminit Maro   *παλίουρος*
'Carduus, et spinis surgit Paliurus acutis'                  *Ib Cnoth leana*
Tilea etc                                                    *Vide Plin. L13, c.19.*
Hisp. Xara. Plin lada 1.26 c.8[581]
Vide cisthos apud eund 1.24 c.10
Alvarquoque[582] Persicum praecoquum. Nebr. est arbor in Ibernia sativa ferens fructum hunc. apud eum vide Persicum.

---

574.   'Rosariae' deleted before Lapilli, vel should then be left out.
575.   After 'busca?' fortasse hec in a different hand deleted, and after that an illegible word deleted.
576.   G. Leuce deleted and λεύκη inserted.
577.   G. Daphne deleted and δάφνη inserted.
578.   G. ?Ame deleted and ἀκτῆ ἀκτίς inserted.
579.   Inserted from the R. margin. Torno? Mesuilae deleted. Sambucus meyulla Pl L.16, c.25.
580.   In L. margin L. Agrifolium, G. Paliures, H. Azelo, I. Cuilin ?Felog, all deleted.
581.   Crossed out 'readeagh' (and above agh ach is deleted), railegach.
582.   Before 'Alvarquoque' alvaroque? deleted.

Box

The completely wild box or oleastrum (brean corabhon – type of box, Plin. 16, c.16). Let us not pass over the box that is so heavy that it sinks in water and so hard that it does not undergo decay. From it in Ireland they used to make rosary beads.

Poplar

Let the white and the black and the Libian poplar follow (see Plin. 16, c.23, also 24).[583] The leaves of the poplar are on a small trembling little stem and rustling among themselves. There are three types of poplar.

Laurel

Let us add the laurel which has a perpetual green foliage sending out its berries is included in the count.

Elder
Pl. 16, 25.[584]

The elder has large leaves with odoriferous flowers and berries which while maturing are red but when mature are black. The elder has a white spongy pith and a thick bark, grey in colour; it is distinguished by many nodes.

Christ's thorn has berries and thorny leaves as Maro remembers:   Thorn[585]
*The thistle and the thorn bush rise up with sharp thorns.*[586] Tilea etc. Hisp. Xara, Plin. LADA Book 26 c8.
See cisthos in the same author 24, c.10., rock rose.
Also Alvar, early ripening peach. See Nebrissa. There is a tree planted in Ireland bearing this fruit. In Nebrissa see Persicum.[587]

583.   Pliny, book 16, chapter 35 section 85, and chapter 36 section 87.
584.   Pliny, 16. c.18 section 30; 17.c.20 section 34.
585.   Pliny, 24. c.13 section 71; 16. c.30 section 53.
586.   Virg. *Ecl.* 5,39.
587.   Peach tree.

# C. XLV
## Frumenta[588]

*Milium croneacht muilin an?*

Arborum Iberniae copiosiorem mentionem ut alii faciant fore spero. Hoc a me dicta esse sufficiat et de sylvestribus, et de frugiferis. Ad aliam frugum speciem, frumentum transibo. Fruges autem illud omne quod ex fructibus terrae in alimoniam vertitur, dicitur: frumentum. Vero fruges, ex qua panis confici solet, appellatur.

*Frumenta, vide dilligenter Plin. L.17 c.7. Croineachtmaol et maolroban.*

In Ibernia frumentorum omnium optimum est triticum: cuius et delicias siliginem ibidem conspicies.

*1. L. Triticum*
*G. πυρος*[589]
*H. Trigo*
*I. Crothneach*

(croineacht habet aristas far Fol 45.
Croineacht moal caret aristis et est mutica spica, Triticum. Plin 18.7&8.Croneach fhrancach, est magnum et coloris magis caesii)
Sequuntur alia, ut tipha, quam novi sigalum vocant:

*2. L. Tipa*
*G. τίφη*[590]
*H. Zenteno*
*I. Siagail*

ut hordeum notissimum frumentum ex quo potissimum cervisia potus genus fit: ut avena.

*3. L. Hordeum*
*G. Κριθή*[591]
*H. Zevada*
*Ib. Orna*

*4. L. Avena*
*G. ἀιγίλωψ, βρόμος*[592]
*H. Avena*
*Ib koirke (coirce)*
*Vide diligenter Pl.*
*28.c7 avena sativa*
*coirce avena non*
*sativa quam Plin.*
*vocat vitium vitium*
*coirke sealein*

---

588. Before 'frumenta' Iberniae is deleted.
589. G. Pyros deleted and πυρος inserted.
590. G. Tipa deleted and τίφη inserted.
591. G. Crithe deleted.
592. G. Egiliops deleted and ἀιγίλωψ, βρόμος inserted.

# C. XLV
## Grains

I hope that others may make a more extensive mention of the trees of Ireland. What has been said by me may suffice concerning the woodland and fruit bearing trees. I will pass on to another type of crop – grain. Everything that is turned into nourishment from the fruits of the earth is called produce.

*Check Pliny diligently 18,7. Croineacht gruagac (hairy wheat, ordinary wheat for bread). Croineacht maol (buck wheat) and maolroban (cruithneacht) Milium croineacht ?muilin (milling wheat)*

The product from which bread is usually made is called grain. In Ireland wheat is the best of all the grains, whose delight, wheaten flour, you will see there also.

Wheat is a bearded grain, fol. 45. Croineach Maol (buck wheat is lacking the beards and does not have spines. Triticum Pl. 18, 7 & 8. French wheat is large and in colour more blueish grey.) *Wheat*

Others follow such as tipha (one grained wheat) which modern-day people call rye. *Rye*

Barley is the most well known grain from which mostly a type of drink, beer, is made: also oats. *Barley*

*Oats[593]
See Pl. 23.c7 for cultivated oats. Pliny calls wild oats 'vitium'.
See Plin 18,17*

---

593.　O'Sullivan Beare does not say anything in the text concerning oats.

# C. XLVI[594]
## Legumina[595]

Ad haec frugum genus aliud, legumen accedat quod ex terrae
satis in siliquis nascitur.

*1. L. Faba*        ut faba inter frumenta etiam relata, quod ex ea panis
*G. ἐρέγμος*[596]     nonnunquam conficitur:[597]
*H. Haba*
*I. Poinri*[598] *quaere*
*pisum apud Plin.*

*2. L. Phasiolus*     ut phasiolus, ut pisum ut ervilia, ut cicera ciceri simillimus.[599]
*G. φαδίολος,*
*ἰσόπυρον*[600]

*H. Frisales,*     *Vitium frumenti, plinio lolium vel aera roile, urica roadan, Plin.*
*Aruaja I.*       *18.17.*

*L. Ervilia*      ut ervum omnibus dissimiler.[601]
*H. Aruazaa*
*Pishapaill*

*3. L. Lens*      ut lens utroque minus globosa.
*G. φακος*
*H. Lenteja*
*I. Pisean*

*4. L. Ervum*
*G. ὄροβος*
*H. Yervo*

---

594.   XXXX deleted before XLVI.
595.   Hilium an pismolin deleted.
596.   G. Geraemos deleted and ἐρέγμος inserted.
597.   Inserted from the R. margin: Ervilia., H. Aruaza, Pischapaill, Plin.
598.   I. Poinr deleted, poinri inserted.
599.   Inserted from the R. margin *Pis minus, pisgeal maius est.*
600.   G. Phasiolis deleted and φαδίολος, ἰσόπυρον inserted.
601.   Ut ervum (omnibus dissimiler deleted).

# C. XLVI
## Leguminous Plants

*Broad bean check the pea in Plin. 18.21*

To these may be added another type of produce named legume, which grows in pods from plantings in the earth, such as the broad bean, also mentioned among the grains because bread is sometimes made from it.

*Bean*

*Horse-vetch*

The bean, the pea, the vetch, the chickling-vetch which is very similar to the chick pea.

*The pea is smaller; pis geal is bigger.*
*The fault of the grain in Pliny is the tare or darnel. Is the cankerworm roadán? See Pliny, 18.17.*

*Vetch*

The bitter vetch is different from all others.

*Lentil*

The lentil is less rounded than both, as is the bitter vetch, which is not unlike them all.

# C. XLVII
## Herbae

Cum hic aliquas herbas coniungamus ex innum eris. quas Ibernia in varios humanae vita usus utilissimas procreat.

*1. L. Linum*
*G. λίνον*[602]
*H. Lino*
*I. Llin*

Hic linum seritur herba notissima, quae in lanae millitiem contusa in intimas vestes netur.

Sequitur cannabis funibus aptissima.

*2. L. Canabis*
*G. κάνναβις*
*H. Canamo*
*I. Cnaib*

Sponte nascitur spartum: et iuncus a iungendo dictus, quia ad iuncturas aptus est.[603]

*3. L. Spartum.*
*H. Sparto*[604]
*G. σπάρτιον*[605]

Ferulacei generis frequens est foeniculum ad condienda plurima idoneum.

*5. L. Foeniculum*[606]
*G. μάραθρον*[607]
*H. Hinojo*
*Ib Fineill*

Cucumis herba sarmentis praelongis humi repentibus et in ramos vitis modo sese fundentibus folio rotundo, cucurbitae folio aliquanto minore, flore luteo a fructus oblongi curvitate nomen accepit.

*6. L. Cucumer*
*G. σίκυος, σίκυς*[608]
*H. Cohombro*
*I..*

Cucurbita vero Hortensis herba fructus omnes cum arborum tum herbarum superat foetus magnitudine, de rapum fruticem satis notum.

*7. L. Cucurbita*
*G. σικύα,*
*κολοκύνφη*[609]
*H. Calabaza*
*Ib Calabus et*
*pompina*[610]

---

602. G. Linon deleted and λίνον inserted.
603. Deleted in L.margin L. Spartum, G. Schoinos Bryllon, H. Iunco, I. Luachair.
604. G. Canabis deleted and σπάρτιον inserted.
605. Deleted in R. margin, Spartum, An Bithineach, Luachair? cruach, G. ???, G. ???, ???.
606. There is no number 4.
607. G. Marathron deleted and μάραθρον inserted.
608. G. Sicys deleted and σικυός σίκυς inserted.
609. G. Colocynthe deleted and σικύα κολοκύνθη inserted from the L. margin.
610. There is no number 8.

# C. XLVII
## Grasses (herbage)

To these let us add other grasses from the innumerable ones which Ireland produces, most useful for the various needs of human life.

*Flax*   Here, flax is sown, a very well known grass which is pounded to the softness of wool and woven for underwear.

*Hemp*   Next follows hemp most suitable for ropes.

*Broom*   The broom grows of itself; and the rush, so called from joining because it is good for joints.

*Fennel*   Of the family of giant fennel, fennel is common. It is useful for seasoning many things.[611]

*612, 613*

*Cucumber*   The cucumber plant gets its name from the curvature of its longish fruit, with very long tendrils creeping on the ground, and developing into branches after the manner of a vine. It has a round leaf somewhat smaller than the melon leaf, with a yellow flower.

*Melon*   The garden melon plant exceeds all the fruits of both trees and herbs in the size of its produce. Add the sprouting turnip a plant which is well known. *Calabas et pompina* (meal beachán)

---

611.   Giant fennel in the pith of which sparks of fire keep alight; for which reason it is thought to have been used by Promethus when he stole the fire from heaven, Pl. 13.22.42.

612.   The following are deleted in the L. margin, L. Iunco, G. Schoinois Bryllan, H. Iunco, Luachair.

613.   There is no number 4.

Adde rhapanum nihilo ignotiorem. raphanus vocatur ceso radicula.

*Nebeta, hisp. Nebda.*[615]

9. *L. Raphanus*[614]
G. ῥαφανίς
H. *Rauano*
I. *Rabuin*

Adnumeratur et pastinaca Graecis staphylinos dicta, quae[616] albam habet radicem.[617]

10. *L. Pastinaca*
G. σταφυλῖνος
H. *Zanahoria*
I. *Meaca dearg*[618]

Huic simillimum gingidium sequatur, quod et daucum dicitur.

11. *L. Gingidium*
G. γιγγίδιον[619]
H. *Velesa Bisnaga*
(*Biznaga?*)
I. *Mealuchain*

Et non dissimile siser.[620]

*L. Siser*
G. σίσαρον
H. *Chirisia*
I. *Caisearban*
(*Caisearbhan*)

Accedunt betae, quae quia sunt insipidae, fatuae peculiari epitheto nominantur iuxta illud Martialis.
*Ut sapiant fatuae fabrorum prandia betae*

13. *L. Beta*
G. τευτλίον,[621]
τευτλον, σευτλον
H. *Azel*
I. *Biatas*

---

614. Before L. Raphanus the following are deleted in the R.margin. L. Rapum, Rapulum? Rapsrium, G. Congre, Congrus in the L. margin. ??? and ??? are inserted. H. Nabo, I. Turnaps.
615. Deleted an Iber meaca vide Plin. L.20 c.14 an Ib. meaca bhan examina.
616. Crossed out after 'quae' altera alba, altera rubra est.
617. G. Staphulinos deleted and σταφυλινος inserted, σταφυλινος is more correct.
618. A large blot obliterates the end of dearg.
619. G. Gig but rest deleted by large blot.
620. Inserted from the L. margin huius caiseraban duo genera alterum dulcem alterum amarum radicem habet vide Plin. L.19, c.5
621. G. Tetulum deleted and τευτλίον, τευτλον, σευτλον inserted.

And the radish is no less known, it is called a small cut root.       *Radish*

Let us count in the carrot, called *staphilinos* by the Greeks,       *Carrot*
which has a white root.[622]       *Check this. Pl. 20, 14*
       *I. Meaca bhan*

After this comes the almost identical French carrot which is       *French carrot*[624]
also called *daucum*[623] (a parsnip-type plant).       *Check this.*

And not dissimilar is the dandelion.       *Dandelion*

*Of this dandelion, among the Irish there are two types: one has a*
*sweet root, the other bitter. Pl. 19, 5.*

Then comes the beet which, because it is tasteless, is called by a       *Beet*
peculiar name insipid, according to Martial:[625]
'The artisans' dinner tastes of the insipid beet.'

---

622.   The manuscript initially read 'one sort has a white root, and another a
       red root', but this has been erased in favour of the text given above.
623.   Daucum, a parsnip-type plant. Pl. 19.5.27
624.   Only gig seen, rest is covered by a blotch.
625.   Martial *Ep.* vol. 2,13,13. Correct Spanish name is *rabwand*.

Coepa notissima magno capite non est silentio praetereunda.

*14. L. Coepa*
*G. κρόμμυον*[626]
*H. zebolla*
*I. Inniun*

Neque porrum tacendum alterum sativum, alterum capitatum.

*15. L. Porrum*
*G. πράσον*[627]
*H. Puerro*
*I. Iluis*

*16. L. Allium*
*G. σκορόδιον*
*σκόρδον*[628]

Emorandum allium cuius et vis et utilitas magna celebratur.

*17. L. Lectuca*
*G. Didrax*[629]
*H. Lechuga*
*Ib Leitius*[630]

Lactucam non est quid longius explicemus, et in Ibernia notam, et Martialis carmine insignem.
*Claudere quae coenas lactuca solebat anorum*
*Dic mihi cur nostras inchoat illa dapes?*

*18. L. Brassica*
*G. κράμβη*[631]
*H. Verzo, repollo*
*I. Coil, cabaisti*

Brassica hic non deest, cuius laudes exequi, longum esse, Plinius secundus est author, cum Chrysippus medicus volumen ei dicaverit per singula hominis membra digestum. Eam in tres species veteres divisere crispam[632] latisque foliis a caule exeuntibus Seam; et tenuioribus simplicibus, atque densissimis foliis amariorem, quae Crambe vocatur.

*19. L. Nasturtium*
*G. κάρδαμον,*
*καρδαμίς*[634]
*H. Nastuerzo berros,*
*I. Pubuiriagais builta*

Nasturtium a tormento nasi[633] quem mordacitate quadam cruciat, dictum, duplici genere hic reperitur, alterum hortense, alterum sylvestre, quod et aquaticum nominator, et laver et Graecis sion.

---

626.   G. Crommyum deleted and κρόμμυον inserted.
627.   G. Prason deleted and πράσον inserted.
628.   G. Scorordor is deleted in L. margin and the Greek word inserted from the R. margin.
629.   Didrax is one of the few Greek words that have not been converted to Greek lettering.
630.   Deleted in the L. margin: L. Beta, G. Teutlon, H. Azelga, I. Bielis.
631.   G. Crambe deleted and κράμβη inserted.
632.   'Eam' deleted after 'crispam'.
633.   'Nasi' deleted before 'tormento'.
634.   G. Cardamon deleted and κάρδαμον, καρδαμίς inserted.

The onion, best known for its great head, must not be passed *Onion* over in silence.

Nor must one stay silent about the leek, one grows underground, *Leek* the other overground.

*Garlic*    Garlic must be remembered, whose great power and usefulness is celebrated.

*Lettuce*    There is no need to explain further the lettuce, both well known in Ireland and famous from a poem of Martial:
*'Tell me why is it that lettuce which used to end our grandfathers' meals, begins our banquets.'*[635]

*Cabbage*    The cabbage is not lacking here. Pliny the Younger is the authority that its praises would take a long time to complete. Since Chrysippus, a doctor, dedicated to it a volume encompassing every part of the human body. The ancients divided it into three species, the crisp type with broad leaves arising from the stalk called the Seam; the more bitter type with more slender simple and very dense leaves which is called Crambe.

*Cress*    Cress is named from an irritation of the nose which it torments with a certain tendency to sting. Here, it is found in two types: one the garden type, the second a woodland type which is called watercress, and *laver,* and *sion* by the Greeks.

---

635.   Martial *Ep.* vol. 2, 13,14.

*20. L. Origanum*
*G. ὀρίγανον*[636]
*H. Origana*
*I. Origanum*[637]

Origanum odoris gratissimi suavitate memorandum est, cuiusflores exalbo purpurei umbella supra cacumen comante emicant.

*21. L. Mentha*
*G. ἡδύσμός*[638]
*H. Yerba Buena*
*Ib Miuntius, mismin*

Mentha odore suavi aestate viret; hyeme flavescit: eius sylvestre genus mentastrum vocatur.

Accedit malva satis nota cuius genera duo meminit Plinius Malapon maiorem, et alteram malachon.[639]

*22. L. Malva*
*G. μαλάχη*[640]
*H. Malva*
*I Hocuis, leamhan*
*Vide malache,*
*columellae*

Carduus foliis paucis, atque spinosis, muricatisque cacuminibus surgit. Eius duo sunt genera; alterum fruticosius, alterum unicaule crassius, quod Graeci scolymon dicunt, florem purpureum mittens inter medios aculeos celeriter canescentem et cum aura abeuntem.

*23. L. Carduus*
*G. σκόλυμος*
*H. Cardo*
*I. Hardechor*[641]

Ruta luminibus claritatem adfert, venerem coercet.

*24. L. Ruta*
*G. πήγανον*[642]
*H. Ruda*
*I. Rut*[643]

Effodiuntur hic et radices longae, quibus aqua coctis panni rubro colore inficiuntur, quaere nomen Latinum et Graecum.

*25. I. Daith duimhi*[64]
*Priamh gdhaithi*[645]

---

636.   G. Origanos deleted and ὀρίγανον, ὀρίγανον inserted.
637.   Deleted in L. margin 'Vide canila bula apud Plin, delacambum L.20 c.15, 16. Vide arist hist – the rest is illegible.
638.   G. Hedusmos deleted and ἡδύσμός inserted.
639.   Inserted from the Right margin Hermonium1.7, asprenum Pl. L.27, C.V Deleted 'Hemina iber. Hiem herba odifera para? habet florem albam'.
640.   G. Scolymos deleted and μαλάχη inserted.
641.   *Hardechoc* sounds like 'artichoke'.
642.   G. Peganun deleted and πήγανον inserted.
643.   'Routh' deleted.
644.   Inserted from the L. margin – Aleliria? I. Samroig deleted, de trifoliis deleted, generibus videplinium deleted, trifolii species acleosior aleluia nonullis dicitur.
645.   Spelling may be incorrect, probably dhaithi.

Oregano must be recorded for the sweetness of its very pleasant scent whose purple flowers shine out from an umbrella which is shaggy on top. *See cunilabula in Pliny and Dalecampius, book 20, c.15 and 16. See Aristotle, book 9, History of Animals, 6 and 27.*

Mint grows strong in summer with a sweet smell. In the winter it goes yellow, its wild type is called *mentastrum*.

Next comes the mallow, well known, of which Pliny mentions *Mallow* two types: the larger *malapen,* the other *malachen* (type of mallow). *See malache of Columella. Herminoum et asperum Pl. L.25 c.v. Anise or estin? in Ireland – anis? See Plin., L.2 c.17.*

The thistle has few and spiky leaves; it rises up with purple tops. *Thistle* Of those there are two types, one of which is bushy, the other thicker with a single stalk and which the Greeks call *scolymon,* sending out a purple flower among the central thorns, quickly becoming white and blowing away with the wind.

Rue brings clarity to the eyes and restrains sexual impulses.     *Rue*

Long radishes are also dug up here, which when cooked in water dye the clothes red. Check the Latin and Greek name.

Trifolium a tribus foliis, quibus constat dictum, hic abundat, album florem producens.

26. L. Trifolium
G. τρίφυλλον[646]
H. Trebol
I. Seamroig seamur

Thymus frutex humilis odoratui gratissimus purpureum generat florem.

27. L. Thymus
G. θύμον[647]
H. Tomillo
I. Toim

Gramen herba satis nota, quae geniculatis inter nodiis serpens. Novas radices crebro spargit, iumentis gratissimum pabulum

28. L Gramen
G. ἄγρωστις[648]
hic optimum est.
H. Gramma
I. Irthinn

De fungis, sive boletis quos Iberni nihili faciunt, non est quid fusius dicamus. Dic ex Plin de fungis arborum.

29. L. Fungus
G. Myces, μύκης
H. Hungo
I. Pucapeill

30. L. Absinthium[649]
G. ἀψίνθιον ἄψινθος
H. Axenjos
I. Mor motir

Neque absynthium longa explicatione indiget, herba perquam amara, sed ad multa salubris.

31. L. Artemisia[650]
G. ἀρτεμισία
H. Artemisia
I Hensae et luis nabhrainc

Artemesia absynthii modo fruticosa, foliis maioribus, atque pinguioribus duplex est, altera latioribus, altera minoribus, cum antea. Parthenis nuncuparetur. ab Artemisia regina nomen accepisse fertur.

32. L. Plantago
G. ἀρνόγλωσσον[651]
H. Llanten?
I. Cuach phadric

Plantaginis hic due genera alterum minus, angustioribus, atque nigrioribus foliis, linguae pecorum simillimus, caule anguloso, et in terram inclinato in pratis nascens; alterum maius foliis cortarum modo inolusum, quae quia septena habet, a nonnullis heptapleuron.

---

646.   G. Triphullon deleted and τριφύλλον inserted.
647.   G. Thymos deleted and θύμον inserted.
648.   G. Agrostos deleted and ἄγρωστις inserted.
649.   ἀψίνφιον ἄψινθον inserted Abhsynthium deleted.
650.   ἀρτεμισία inserted Artemisia deleted.
651.   G. inserted ἀρνόγλωσσον Arnaglossa deleted.

The shamrock is called from the three leaves of which it is *Shamrock* composed. It is abundant here, producing a white flower.

Thyme, a low-growing plant, very pleasant to smell, puts forth *Thyme* a purple flower.

Couch grass is a plant well known, of creeping habit with *Grass* jointing internodes, it sprouts new roots repeatedly. Here it is an excellent fodder, very pleasing to beasts of burden.

Concerning fungi or mushrooms which the Irish make nothing *Mushroom* of, there is no reason why we should discuss further. Tell about tree funghi from Pliny.

*ormwood* Nor does wormwood need a long explanation. It is a very bitter plant but healthgiving for many conditions.

*ugwort* Mugwort is bushy like wormwood, with larger leaves and juicier; there are two types, one with broader leaves, another smaller. Since formerly it was called Parthenis, it is held to have got its name from Queen Artemisia.[652]

*antain* Here, there are two types of plantains, one smaller with narrower and blacker leaves, very like the tongue of cattle, with an angled low-growing stalk sprouting in meadows; the other type larger, covered with leaves like ribs; because it has seven at a time it is called *heptapleuron* (seven-ribbed) by some.

---

652.   The wife of King Mausolus of Caria to whom after his death she built a celebrated mausoleum.

*33. L. Buglossus*
*G. βούγλωσσος*
*βούγλωσσον*[653]
*H. Borrajas*
*I. Borraiste*

Buglosso boum linguis simili eam praecipuam vim inesse Plinius tradit, ut in vinum deiecta animi voluptates augeat.

*34. L. Salvia*
*G. ἐλελίσφακον*[654]
*H. Salvia*
*I. Saiste*

Salviam praeterire minime putavi (quae frutex est satis notus) inde dictam (sicuti traditur) quod multis humanae vitae[655] incommodis opituletur, homines salvos et incolumes reddendo. Illa habet virgas quadrangulares, folia aspera, crassa, incana, odore quidem gravi, non tamen ingrato.

Hyssopus herba notissima, pulmoni valde utilis; hyssopum quoque nominata hic provenit.

*35. L. Hyssopus*
*G. ὕσσωπος*[656]
*H. Hyssopo*
*I. Hysoi?*

Helleborus etiam veratrum vocatur: insignem vim purgandi habet.
Eius duo sunt genera, albus vomitione, niger alvi exoneratione causas morborum expellit.
   *Huius duo genera opud celsum nigrum et albon.*

*36. L. Veratum*
*G. ἐλλέβορος*[657]
*H. Vedegambre*[658]

Rosmarinus memorandus, et odore et salubritate commendabilis.

*37. L. Rosmarinus*
*G. λιβανωτός*[659]
*H. Romero*
*I. Rosmarinum*

Sinapis effectu ignea, sapore acerrima semen fert in hyberna adcondimenta utilissimum, tanta acrimonia ut comedenti lachrymas cieat.

*38. L. Sinapis*
*G. σίνηπι*
*H. Mostaza*
*I. Mostaird*

---

653.   Buglossus deleted.
654.   Eleisphakon deleted.
655.   After vitae a second 'vitae' is deleted.
656.   G. Hyssops deleted and ὕσσωπιν ὕσσωπινς inserted.
657.   Helleboros deleted and ἐλλέβορος inserted.
658.   I Meaca bui deleted.
659.   A blot obscures what was deleted, λιβανώτος inserted.

*rage*[660]

Pliny tells the story that there is a special power in borage (which is similar to the tongues of cows), so that when put into wine it increases the pleasures of the mind.

*ge*

I thought that I should not pass over sage (which is a well-known shrub). It is named from this (as tradition has it) that it may help many misfortunes of human life by rendering men safe and sound. It has quadrangular twigs, thick, hard, greyish leaves with a certain heavy smell, however not unpleasant.

Hyssop is a well known herb very useful for lungs, it grows here. *Hyssop*[661]
It is also called Hyssopum.

Hellebore is also called *veratrum,* it has outstanding power of *Hellebore* cleansing the bowels. It has two types; the white one drives out the causes of disease through vomiting, the black one through lightening the belly. *According to Celsius, there are two types: black and white.*

Rosemary needs to be remembered. It is recommended for its *Rosemary* odour and wholesomeness.

The seed of the mustard, with its fiery effect and sharp taste, is *Mustard* most useful in the winter for seasoning. It has such sharpness that it drives out tears from the person eating.

---

660.  Boglossus means cow's tongue.
661.  Hyssopos deleted.

Acetaria, quae acetum nonnihil sapit, in Ibernia notissima
addatur.

39. *L. Acetaria*
*H. Azedera*
*I. Samhbha*
*Vide lapathus*
*apud Plin.*

*L. Hedera*[662]
*G.* κισσός
*H. Yedra*
*Ib. Einnean*

Hedera cum se sustinere non possit, muris et arboribus innixa
surgit. Vide infra helix.
infra Helix.

*L. Lappa*[663]
*G.* ἄρκειον, προσωπίς,
προσώπιον
*H. Amor de
Hortelano*[664]
*I. Leadan liosta*

Lappa capite leviter hammato pilosis rebus haerens triplex est.
Maxima latinis personata, graecis arcion dicta, iuxta. Flumina
rivosque crescit. Minor quae φιλάνθρωπος et aparine vocatur,
hominum vestibus haeret, κύνωψ minima canaria canibus affigitur.

*Vide philanthropos*
*apud Plin L. 24 c. 1*

*39. L. Serratula*
*G.* κέστρον, ψυχότφον
*H. Betonica*
*I. beathbhuin*

Betonica, sive vetonica et serratula in multos usus commendatur.

*40. L. Apium*[665]
*G.* σέλινον
*H. Peregil*
*I. Persil (persil)*

Apium usitatum iunge.

*Examina*

*L. Rhoea*
*Ib. Amapola*
*Ib. Cogail*
*(Plin. 19, 6, 8)*

---

662.  Hedera and its description is inserted from the L. margin. Hedera and
      lappa are not numbered.
663.  Lappa and its description are inserted from the L. margin.
      H. Amor de Hortelano and I. Leadain Hosta are inserted from the R.
      margin using an insertion mark.
664.  *Amor de hortelano* means 'gardener's delight'.
665.  Before Apium Petr is deleted. In the text 'apium' is inserted above the
      line and 'petrapium sive peiraxilium in condimentis' is deleted before
      'usitatum'.

Let us add sorrel, which tastes somewhat of vinegar, and is well known in Ireland. *Check Lapathus in Pliny.*[666]

*Sorrel. See lapathus in Pliny*

Ivy is not able to sustain itself on its own; it grows up supported by walls and trees. See below *helix*.[667]

*y*

The burr is divided threefold, sticking lightly with a hooked head to hairy material. The largest is called by the Latins *the largest*, by the Greeks it is called *arcion*. It grows next to rivers and streams. There is a smaller one, which is called *philanthropos*[668] and *aparine*: it sticks to the clothing of men. Κυνώψ,[669] the smallest, *canaria* gets stuck on dogs.

*urr*

*See philanthropos in Pliny book 24, chapter 17*

Betany (or vetonica or serratula) is commended for many uses.

*etany*

Add that parsley is used.

*arsley*

*ild poppy*
*Plin. 19, 6, 8)*

---

666.  P1.20, 21, 85 § 231
667.  i.e. a type of ivy, in architecture the small volute or ornament under the abacus of a corinthian column representing the tendrils of the ivy.
668.  *Philanthropos* means a lover of man.
669.  κυνώψ means 'dog-eyed'.

41. L. *Petroselinum*
*Petroselinum*
G. πετροσέλινον[670]
H. *Peregil de mar*
I. *Creirc*

Neque Petroselinum in petris optime proveniens praeteribis, semine nigro, sapore amaro.

[671]

42. L. *Alga*
G. φῦκος
H. *Ovas*[673]
I. *Raibh Locha*

Alga herba in aquis marinis nascens, crassa foliis ex parte G. φῦκος supernatantibus nantium pedes saepe alligat.[672]

43. L. *Urtica*
G. ἀκαλήφη
ἀκαλύφη[674]
H. *Hortiga*
I. *Neantoig*[675]

Urticam ab urendo dictam, quod eius folia tacta, non aliter, quam si urerent, loedunt, ne silentio involvamus.

44. L. *Gladiolus*
G. ξιφίον[676]
H. *Espanada*
I.[677]

Gladiolus folia habet gladii figuram referentia

45. L. *Felix*
G. πτερία, πτερίς[678]
H. *Helecho*
I. *Raithneach*

Filicem hic multam videbis, quam sine flore, et semine esse, scriptores affirmant. Ex eius cinere vitrum fit.

*Filicus genera apud Plin. L.27 C.9, fuse.*

46. L. *Lupus*[679]
G.
H. *Hombrezillos*
I. *Hoip (Hopa)*

Ne praetereatur lupus quo cervisia conditur.

670.  G. Petrosilulon deleted and πετροσέλινον inserted.
671.  Inserted from the R. margin – Rhoea Hisp. amapola 16 cogal Plin. L.19, c.18.
672.  This paragraph is surrounded on three sides by a continuous line.
673.  Deleted – I Raibh locha, feamanach. Seven other illegible words deleted.
674.  G. Acalyphi deleted and ἀκαλήφη ἀκαλύφη inserted.
675.  Deleted in the L. margin below 'neantoig' – 'Gloriem vide iris'. A large blot obscures the first words of the next line then ? aroros et aphrodisias.
676.  G. Xiphion deleted ξιφίον inserted.
677.  Deleted and illegible.
678.  G. Pteris deleted and πτερία πτερίς inserted.
679.  Inserted in a different hand in the left margin. I. Hoip, H. Hombresillos.

ock parsley

Nor will you pass by the rock parsley, growing best in rock; it has a black seed and a bitter taste.

eaweed

Seaweed: a grass growing in sea water, it is dense with the leaves partly floating on top. It often entangles the legs of swimmers.

ettle

The nettle is called from the fact that it burns because the leaves cause injury when touched as if they were burning; let us not wrap it up in silence.

ladiolus

The gladiolus has leaves recalling the shape of a sword.

ern[680]

The fern you will see a lot here. Writers affirm that it is without flower and seed. From its ash, glass is made.

lop

Let us not pass over the hop from which beer is made.

---

680.   Marginal, the types of fern in Pliny I, 27,9 widely dealt with.

# C. XLVIII
## Flores

Herbarum quas retulimus, nonnullae florem gignunt, in alios tamen usus utiliores: quibus et alios flores addam.

*1. L. Rosa*
*G. ῥόδον*[681]
*H. Rosa*
*I. Ros*[682]

Inter omnes primum locum rosae deferamus. Ea flos est e spina nascens, in rubo quoque proveniens, cortice primum granoso inducta germinat, mox intumescente, et in virides alabastros, sive papillatos corymbos fastigiato, paulatim rubescens, dehiscit, ac sese pandit: in sui calicis medio luteos apices stantes complectitur.[683]

Ad rosam proxime accedit nobilitate lilium, nulli florum excelsitate secundum, collo languido, capitis oneri vix sufficiente, eximio candore, foliis foris striatis, et ab angustiis in latitudinem sese paulatim laxantibus, offigio calathi, labris per ambitum resupinis.

*2. L. Lilium*
*G. κρίνον*[684]
*H. Azuzena*
*I. Lil*

*3. L. Viola*
*G. ἴον ιονανθός*[685]
*H. Violeta*
*Ib Sailcuach*

Inde viola sequitir, cuius planta folium hederae non dissimile, nigrius tamen, atque tenuius et cauliculum a media radice prosilientem, floremque vel album, vel purpureum, vel luteum fert.

686

*4. L. Solaris herba*
*G. ἡλιοτρόπιον*
*ἡλιοτρόπον ἡλιοπους*
*ἡλιοσκοπιον*
*ἡλιοσκόπιος*
*τσίκοκκον*

Memoratu haud indignam ducimus[687] ob miraculum eam herbam: quae ad varios solis orbem lustrantis aspectus die, etiam nubilo sese circumagit, sydus intuens, calice penitus

---

681. G. Rhodon deleted and ῥόδον inserted.
682. Deleted from the L. margin: Scarragdhig, ros muchuir et miuc or fructus.
683. Inserted from the R. margin – Tribulg – Cynobatos – abrojo vide suis logis et esscoramujo. Rhoea, Hisp. Amopola – Ib cogar vide Plin. L.17 c.6 et ad finem.
684. G. Crinon deleted and κρίνον inserted.
685. G. Ion deleted and ἴον ιονανθός inserted.
686. This whole paragraph is inserted from the R. margin of page 60 of the MS which is a whole page after the names which are in the L. margin of page 59R.
687. 'Duce' deleted before 'ducimus'.

# C. XLVIII
## Flowers

Of the plants which I have recounted, some produce flowers but they are more useful for other purposes. To these I shall add other flowers.

Among all flowers let us give first place to the rose. This is a *Rose* flower arising from a thorn bush coming forth also from a *Tribulg, cynaobatos,* bramble bush. It sprouts forth covered over first with grainy *abrojo: see in their own* cortex, which soon swells insize into a point and into green *place; also escoramujo.* rosebuds or budding clusters; in a while growing red it opens up and spreads itself out. In the middle of its own calyx it embraces the standing yellow tips.

The lily approaches closest to the rose in nobility, second in *Lily* height to no other of the flowers, it has a weak neck scarcely strong enough for the weight of the head. It is an outstanding brilliant white, with leaves striated on the outside with ridges which from being narrow, widen gradually in breadth like a wicker basket with lips bent back round the edge.

*iolet* Then follows the violet, a plant whose leaf is not dissimilar to the ivy in its foliage, but blacker and more slender and it produces a little stalk springing forth from the centre of the root, and a flower either white or purple, or mud-coloured.

*unflower*[688] We consider the sunflower well worth remembering because it is a wonder which even on a cloudy day moves itself around to the various aspects of the sun shining on the world, looking attentively at the star, its cup completely expanded as though captured by love of it; by night indeed as if affected by the desire for light, it covers over its flower, which may be either white or yellow, by again contracting its cup. The Greeks, aptly, call it *Ἡλιοτρόπιον* since ηελιοζ means the sun, τροποζ a turn.

---

688.   O'Sullivan Beare wonders if nóinín (a daisy) is the sun flower.

(*cont.*) expasso, velut eius amore capta; noctu vero, sicuti desiderio luminis affecta,[689] florem suum sive candidum, sive fulvum occulit, calato rursus contracto. Ἡλιοτρόπιον dixere Graeci, non inepte, cum ἥλιος solem τρόπος mutationem significet. Solarem herbam Latini nomen edunt. Eius esse duo genera. Holioscopium et Tricoccum: illudque altius, et in pingui, cultoque solo crescere, hoc vero in inculto quoque nasci, Plinius author est. Utroque frequento Ibernia est odorata.

| | |
|---|---|
| *4. L. Humile malum*<br>*G. χαμαίμηλον*[690]<br>*H. Manzanilla*<br>*I. Comain miull*<br>*Vide fuse Plin. et*<br>*Ga1.1.3. c.9* | Chamemllon Graecis, Latinis humile malum a mali odore vocatur: Anthemis quoque dicta folio viridia fundit foeniculi foliis similia, et florem album et calices medio pallidum apicem gignit. / florem[691] vero foliis vel candidum,[692] vel melinum vel purpureum[693] in calicis medio pallidum a picem cingentibus gignut. Aliis nominibus eam Plinius vocat Leucanthendia, Leucanthemum, Cranthemon, chamaemolon, Metanthemon. |
| *5. L. Crocus*<br>*G. κρόκον, κρόκος*<br>*H. Azafrán*<br>*I. Croich*[694] | Quibus adnumera crocum magis superioribus temporibus, quam hodie in Ibernia coli solitum. Quid referam plures flores tot in Ibernia hortos, atque campos ornant, ut eorum nomina comprehendere, longissimum fuisset. Unde de cadici posset:<br>*Tot fuerunt illic, quot habet natura colores*<br>*Pictaque dissimili flore nitebat humus*[695] |

---

689.　'Capta' deleted before 'affecta'.
690.　G. χαμαίμελον deleted and χαμαίμηλον inserted.
691.　From florem to end of paragraph is inserted from the R. margin.
692.　'Album' is deleted before 'candidum'.
693.　After 'purpureum' an illegible word is deleted.
694.　Inserted in the L. margin. 'Bog luachair habet magnam medullat? (should be medullam)' Cruaich luachair habet parvam medullam birineach habet nullam medullam. Above 'bog luachair' three lines are deleted: Iris hic? Birineach cruaigh nonin air? Vide spartum vide funus.
695.　Deleted from the R. margin. L. Crocus, G. Crocos, H. Azafran, I. Croic, quibus odde crocum superiuribus, temporibus quam hodie in Ibernia magis coli solitum.

The Latins give the name sun flower. There are two types of this plant, *helioscopium* and *tricoccum*:[696] Pliny says that the first one taller grows in a fertile and cultivated soil and the second in uncultivated soil also. Ireland is frequently scented by both.

*Camomile*

Camomile is called *chamaimelon* by the Greeks and by the Latins the ground apple from the odour of an apple. It is also called *anthemis*. It produces green leaves similar to the leaves of fennel. It sprouts a white flower and it has a pale tip in the centre of the calyx.

*It produces a white or quince yellow or purple flower with leaves surrounding it. Pliny calls it by other names, Leucanthenida, Leucanthemum, Cranthemon, chamaemolon, metanthemon.*[697]

*Crocus*

Number with these the crocus: it tended to be cultivated more in previous times than today in Ireland. Why should I recall more? So many flowers adorn the gardens and fields in Ireland that to include their names would be tiresome. From which it is possible to say of it: *'So many colours were there, as many as nature has; and the earth was shining, painted with various flowers'.*[698]

*The bullrush has a large pith. The hard rush has a small pith and the hardy sea shore rush has no pith.*

With these I thought I would put the last touch of my review of plants. For at present I am not working as a botanist, that I should

---

696.   τρίκοκκος – three berries or grains, a type of sunflower.
697.   Plin. 27, 12, 98 § 124
698.   *Ovid Fasti* IV. 29–430.

Quibus herbarum reptuationi summam manum imponere putavi. Non enim herbarii munere in praesentia fungor, ut[699] herbarum Iberniae exactissimam[700] doctrinam tradam, vel omnia nomina percurram. Sed plurimas[701] praeteriens aliquas nominavi, quae si solae fuissent, eam insulam maxime decorarent.[702] Cui paucas memorare videbor, equidem aequo anima feram, ut is laborem meam damnet, modo me non conviciis carpat, sed patriam laudibus accumulet, eius plures herbas, pluraque alia ornamenta recensendo:[703] Ad quod a me dicta illi impedimento non sunt, et dicenda auxilio erunt.

## C. XLIX
### Sal. nitrum

*Ex aqua genita*

Itaque quandoquidem de his quae ex terra nascentia surgunt, egimus, nunc aliqua ex aqua facta breviter litteris mandemus.

*1. L. Sal*
*G. ἅλς*[704]
*H. Sal*
*I. Sailinn*

Sal, quo in condiendis cibis utimur, ex humore, qui in maritimis rupibus solis ardore siccatur, hic perquam parce invenitur. (Nec ex aqua marina lebetibus ad ignem multus condensatur.)[705]

*2. L. Nitrum*
*G. νίτρον*[706]
*H. Salime*
*I. Salpiter*

In sali simillimam petram solis quoque ardor hic excoquit in nitrariis nimias aquas, quae per aestatem prolixiore pluvia telluri infunduntur. Haec vero petra nihil salsi saporis nihil gelidi rigoris habet, caliditate durare, nubiloso aere fluere, atque liquescere solet. A nitria Aegypti regione, in qua copiosissime provenit, nitrum appellatur.

---

699.  Deleted after 'ut'? exactius picta.
700.  Exactissimam is inserted above the line.
701.  He changes 'pluras' to 'plurimas'.
702.  'Ornarent' is deleted and 'decorarent' is inserted above the line.
703.  'Repentendo' deleted 'recensendo' inserted after it.
704.  G. Hals deleted and ἅλς inserted.
705.  The words in brackets are inserted from the R. margin, with an insertion mark.
706.  Nitron deleted and νίτρον inserted.

hand on a very exact knowledge of the plants of Ireland or that I should run through all the names. But missing out very many, I have named some, which if they were the only ones would have adorned that island to the greatest extent. To some person I will appear to record only a few. Still I will be satisfied if he may find fault with my work provided that he does not condemn me with reproofs but heaps the homeland with praises, by reviewing its many plants and its many other types of ornament. To this end, those that have been named by me will not be a hindrance to him and what remains to be said will be a help.

## C. XLIX
### Saltpetre, originating from water

*Things originating from water*[707]

Thus, since we have gone through those things which, when being born, spring from the ground, now let us entrust to writing, briefly, some of the things made from water.

*Salt*

Salt, which we use in the seasoning of food and which is dried from the sea water by the heat of the sun on maritime rocks, here is found in extreme scarcity. Nor is much of it condensed from sea water in coppercauldrons on a fire.

*Saltpetre*

Here also the heat of the sun on to a rock, similar to salt, boils off excess water in saltpetre[708] pits, which through the summer falls on the land as widespread rain. This rock truly has no taste of salt and no rigidity produced by cold; it tends to harden in heat and tends to become fluid in cloudy weather and to liquify. It is from Nitria, a region of Egypt, in which it is most copiously found, that nitrum (saltpetre) gets its name.

---

707.    This marginal is written in another hand in the original manuscript.
708.    Pl. 81, 10.46.

## C. L
### Metalla[709]

*1. L. Aurum*
*G. Chysos*[710]
*H. Oro*
*I. Oir*[711]

De metallis iam dispiciamus: quorum celeberrimum aurum in humani generis perniciem repertum utinam e vita in totum abdicari posset. Illud et si in Ibernia hodie minime eruatur; suis tamen in venis latere et vetus[712] scriptorum authoritas,et recens fama tradit. Tribus fere modis in orbe nostra comparari,[713] aut in fluminum ramentis, aut puteorum scrobibus, aut ruina montium, Plinius est author. Per eius venam in puteis humor fluit, limo frigoribus hybernis in pumicis duritiam crassescente nominatus

*2. Chryscolla*

Chrysocolla: quae etiam in argentariis, aurariis, atque plumbariis metallis reperitur

*3. L. Argentum*
*G. ἀργύριον ἄργυρος*[714]
*H. Plata*
*I. Argid*

secundum quoque insaniam, argentum, hic effoditur, constat.

*4. L. Argentum vivum*
*G. ὑδράργυρος*
*ἀργυροσχύτος*[715]
*H. Azugue*
*I. Argid beo*

Est et liquor argentum colore, aquam fluore referens nomine argentum.

In argentariis metallis minium laudatissimi coloris pigmentum solet reperiri.

*S. L. Minium*
*G. Mitlos*
*H. Bermellon*

Sequitur hic et vena aeris rubro colore metalli, ex quo nunc et etiam olim ante auri, et argenti usum pecunia signabatur, Ovidio teste *Aera dabant olim, melius nunc omen in auro est.*

*6. L. Aes*
*G. χαλκός*[716]
*H. Cobra*
*I. Cupur*

---

709.  'Iberniae' deleted before metalla. 'Metalla' also deleted in the R. margin.
710.  χρυοσς deleted.
711.  Deleted in L. margin L. Aurum, G. Chrysos, χρυοςξ deleted above it. H. Oro/I Oir, Or.
712.  'Vetusticana' has 'ticana' deleted to leave 'vetus'.
713.  An illegible word is deleted and 'comparari' inserted above the line.
714.  G. Argyrium deleted and ἀργύριον inserted.
715.  G. Hydrarguros and Harguroschytos are deleted in L. margin and G. ὑδράργυρος, ἀργυροσχύτος inserted from the R. margin.
716.  G. Chalc is deleted and χαλκός inserted.

# C. L
## Metals

*Gold*

Now let us take a look at metals of which the most famous is gold, found to be the ruin of the human race; would that it were possible to reject it totally from life. Also, it is rarely mined in Ireland today; however, the recognised ancient writers and recent reports state that it is hidden in its veins. Pliny says that usually gold is found in three ways in our earth, in the gravel chips of rivers, or in trenches of pits, and in the collapse of mountains.

*Borax*

Through its vein in trenches flows a liquid in the form of mud which in the cold of winter thickens into the hardness of pumice: it is called borax,[717] which also is found in silver mines, copper mines and lead mines.

*Silver*

It is agreed that a second source of madness, i.e. silver, is also mined here.[718]

*Quicksilver*

There is also a liquid, silver in colour, which flows like water and is called quicksilver (mercury).

In silver mines red lead, *vermilion* tends to be found. It is a *Vermilion* pigment of most excellent colour.

Here follows a vein of copper metal, red in colour, from which *Copper* money was coined, now and even long ago before the use of gold and silver. According to Ovid,[719] *'once people used to give copper coins, nowadays there is a better omen in gold'.*

---

717.   Chrysocolla.
718.   This is not included in the transcript of O'Donnell.
719.   Ovid, *Fasti*, 1.221.

Ad aeris quidem naturam nonnihil, sed ad auris colorem proxime accedit orichalcum in Ibernia fertilibus venis nascens.

7. L. *Orichalum*
G. ὀρείχαλκον[720]
H. *Laton morisco*
I. *finnbruin*

Ibidem ferrum in plurima vitae commoda, et incommoda effoditur.

8. L. *Ferrum*
G. σίδηρος
H. *Hierro*
I. *Ierinn*

Neque plumbum foecunda gleba negat cuius et fodinae stannum.

9. L. *Plembum*
G. μόλυβδος[721]
H. *Plomo*
I. *Luo*

10. L. *Stannum*
G. κασσίτερος[722]
H. *Stanno*
I. *Stáin*

# C. LI
## Sulphur, Alumen, Terrae genera

1. L. *Sulphur*
G. θεάφιον θέειον
H. *Azuge*
I. *Raibh*

(De genetrice metallorum terra nonnihil addam. Sulphur, quod sive terrae genus, sive metallicum corpus esse velis, certe igneae naturae est, hic gignitur, praecipue in Murea insula *Inismureagh* Connachtae adiacente, ubi, et vulcani conspiciuntur.)[723]

Salsugo alumen ad crystalli colorem accedens hic gignitur;

2. L. *Alumen*[724]
G. στυπτηρία[725]
H. *Alumbro*
I. *Alim*

tum terra colore pallida; tum alia terrae genera in lateres tegulas et alios usus plurimos mirificae suppetunt.

3. *Variae terra generae*

---

720.   G. Sideros deleted and ὀρείχαλκον inserted.
721.   G. Molybdus deleted and μόλυβδος inserted.
722.   G. Etasiteros deleted and κασσίτερος inserted.
723.   The words in brackets are inserted from the L. margin.
724.   L. word deleted – terrae genera.
725.   G. Stypteria deleted and στυπτηρία inserted.

Yellow copper ore has some of the character of copper but it approaches very closely in colour to gold. In Ireland it arises in rich veins.

*Yellow copper ore*

In Ireland also, iron is dug out, it is useful for many conveniences and inconveniences of life.

*Iron*

Nor does fertile soil refuse to yield lead. Its mines also produce tin.

*Lead*

*Tin*

## C. LI
### Sulphur, Alum and Types of Earth

*Sulphur*

Let me add something concerning the earth, the begetter of metals. Sulphur, whether you would prefer it to be a type of earth or a metallic substance, truly it is igneous in nature. Here, it is formed mostly in the island of Murea, Inismuireagh, near Connaught, where also fires of the earth are seen.

Alum salt is found here approaching the colour of crystals; sometimes, a pale earth.

*Alum*

*3. Different types of earth.*

Again, other types of wonderful earth are available for bricks and roof tiles and very many other uses.

# C. LII
## Lapide et gemmae

Lapides iam referuntur quorum hic non inutilia genera ad aedes et oppida constituendum suppeditant.

| | |
|---|---|
| *1. L. Marmor* | Marmor quidem solidum, atque durum, frequens hic conspicies. |
| *G. μάρμαρος* | Album nomine Parium, viride Laconicum vocatum et |
| *μάρμαρον*[726] | Luculleum fere atrum, quo Lucullus[727] summopere delectatus |
| *H. Marmol* | fertur.[728] Quin ex Ibernia in Hispaniam et alias regiones marmor |
| *I. Marmuil* | quaestus causa exportari aestas, nostra testatur. |

| | |
|---|---|
| *2. L. Gagates* | Gagates lapis niger, planus, premicosus, levis fragilis ad plurima |
| *G. γαγάτης* | utilis, |
| *H. Azebache* | |
| *Ib. Vide Antipates* | |

| | |
|---|---|
| *3. Iris Barth* | et iris, vel sexagonius, qui soli oppositus coelestis iridis, vel arcus |
| *Angilcum Catalog* | similitudinem facit, hic esse[729] ab authoribus memorantur.[730] |

| | |
|---|---|
| *4. L. Aetites* | Aetitem quoque ab aquilis in nidis, collocari,[731] memoriae |
| *G. ἀετίτης* | proditum est. |

| | |
|---|---|
| *5. L. Pumex* | Pumicem lapidem cavernosum, levem, aquis innatantem ad |
| *G. κίσηρις* | res scabras laevigandas laudatissimum in Iberniae littoribus |
| *H. Piedra esponja*[732] | offendes. |
| *Ib. cloch mhini*[733] | |

---

726. G. Marmaros deleted and μάρμαρος μάρμαρον inserted.
727. 'Romanus' deleted after 'lucullus'.
728. 'Traditur' deleted and 'fertur' inserted above the line.
729. 'Inutilis' deleted and 'esse' inserted above the line.
730. 'Feruntur' deleted and 'memorantur' inserted after it.
731. 'Reporari'? deleted and 'collocari' inserted above the line.
732. H. Piedra Esponja deleted.
733. Illegible word '?mini' deleted before 'mhini'.

# C. LII
## Stones and Gems

*Stones*

Now let us deal with stones; here, useful types are available for building houses and towns.

*Marble*

Marble. You will see solid hard marble here. White marble, Parian by name, a green marble called *Spartan,* and Lucullan, almost black in which Lucullus is said to have taken exceeding delight.[734] But indeed also, our age bears witness to the fact that marble is exported from Ireland to Spain and other regions for profit.

*Jet*

Jet is a black stone, smooth, very shiny, light, brittle and useful for many things.

*Crystal. Cited by Bartholemaeus Anglicus,* De proprietates rerum.

A crystal (antipathes) or six-sided stone, which when placed opposite the sun makes the appearance of a celestial rainbow or arc. That these are found here is recorded by authors.

*Eagle stone*

It has been handed down to memory that the eagle stone was placed in the nests by the eagle.

*Pumice*

You will come across the pumice stone on the shores of Ireland. Hollow, light, it floats in water and is most praiseworthy for smoothing scruffy materials.

---

734.    Marble found on an island of the Nile of which Lucullus was very fond Pl. 32,22 §6.

| | |
|---|---|
| *6. L. Silex*<br>*G. πυρίτης λίθος*[735]<br>*H. Pedernal*<br>*Ib Cloch chin* | Silex, ex quo ignis excutitur plurimus habetur |
| *7. L. Cos*<br>*G. ἀκόνη*<br>*H. Aguzadera*<br>*Ib Cloch lieth* | Cotes quibus ferrum acuitur haud sunt obliviscendae |

Ne lapides, ex quibus adustis gypsum et calx conficitur, praetereamus.

Occurrunt et alii, qui si aqua madefacti fricentur, ex aliis nigrum atramentum, ex aliis ruber, ex aliis viridis color fit. De molarum, mortariorum, tegularumque[736] lapidibus non est quid longius dicamus.

| | |
|---|---|
| *7. L. Crystallus*<br>*G. κρύσταλλος* | Lapides quoque pretiosis, sive gemmas hic esse fama tenet, ut crystallum ut lichnitem: quae gemma in ciconiarum nidis solita reperiri, remissior carbunculus quibusdam dicitur.[737] |
| *L. Lychnites*<br>*G. λυχνίτης λυχνευς* | *Vide cambrensem et Bartholomeum Anglicum* |

*Puxari gith*[738] *ex accipitrum genere inanet in aere alis titubantibus suspenus, vivit muribus culicibus et similibus.*

---

735.   G. Pyrites deleted and πυρίτης λίθος inserted.
736.   'Tegularumque' is inserted above the line.
737.   This description of the kestrel is in a different hand and is a sudden change from the description of stones. The description of birds has been extensively dealt with by the author in C.XII to XXV. There is no new chapter number and no chapter heading. There is no explanation for the return to birds.
738.   Inserted from the R. margin: Puxairi, gith. ex acipitrum genere manet in aere alis titubantibus suspensus, vivit, muribus, culicibus et similibus.

*Flint*

Flint from which fire is struck is held in high regard.

Whet stones with which iron is sharpened must not be forgotten.

*8. Various types of stones.*

Let us not pass over stones from which, when burnt, plaster and lime are made. There are others which become steeped if rubbed with water, from some black ink is made, from others a red colour, from others a green colour. Concerning mill stones, crushing stones and slates there is no need to discuss further.

*Whetstone*

Rumour has it that there are precious stones or gems here, such as crystal and white marble. The garnet is said by some to be cheaper than this gem which is found in storks' nests.

*Gemstone*

A kind of white marble.

*White marble.*
*See Cambrensis*
*and Bartolomeus*
*Anglicus.*[739]

*Puxari gith*

The kestrel, from the family of hawks, remains in the air suspended on fluttering wings; it lives on mice, midges and the like.

---

739.    There is a return to the description of birds, without any heading, and in a different hand to what was previously used.

*Lon iski,*[740] *vide supra ei coturnicis magnitudine color ventri ablus dorsi, niger caeruleis punctis distinctus corpus oblongum*[741] *ad lacus et flumina agit cauda brevi piscandi causa se mercit in flumen vide riparia*[742] *apub Anglum.*

*Buineam liana ὀνοκρόταλος (the bittern)*[743]

Buinnean liana

Meleagridi fere par plumarum specie, sed macra, et esu ingrata, colore cinereo[744] (sic), rostro, cruribus longis.

Vide Calep. et Plinium. flumina et lacus accolit, pisciculis vescitur: quibus e priore ventriculo eiectis pullos educat.[745]

*Creachoig. Hispanice arandato albo et valde scurrilis Vide arist de monedulis minor graculo maoir pica λύκος videsatiria grain catha. Avicula colore fusco canora an Philomela? Neaska, rostro longo fusco vide friguilla L. Fringillago manticula vide etiam annotationem gabharr roth, gabhar deora, minnan eair vide carpimulcus lagopus est in Ibernia genus perdicis aves diomed sunt (est deleted) in Ibernia.*

Coileach feagh. Mileagride paulo minor, colore niger caeruleis punctis distinctus palpebras habet rubris villis vestitas,[746] rostro et cruribus longis, impetu magno volat, alas fortiter quatiendo. Ei[747] duplex caro exterior colore fusca, interior alba, eum esse avium omnium esu sapidissimum atque gratissimum sunt qui putent. Huic ova albida punctis cinereis distincta, humi[748] parit inter herbas, et frondes nullo condito nido, implumes dum pullos[749] ducens pascit, alisque fovet, vescitur frugibus lini granis, glandibus, acinis, avellanis et aliis fructibus. Excludit pullos duodenos saepe, aliquando minus. Sedet in arboribus.

---

740. 'Kearc iski' deleted before 'lon iski'.
741. 'Palimpes' deleted after 'oblongum'.
742. ? Fringa deleted before 'riparia'.
743. 'Kearc iski' deleted before 'lon iski'.
744. 'Cinero' in manuscript incorrect.
745. An illegible deleted line follows 'educat'.
746. After 'vestitas' superciliaque rubra pedibus crossed out.
747. Deleted after 'ei', 'triplex caro extima cutive proxima nigra colore media lutea intima albida'.
748. After 'humi' 'nidifi' deleted.
749. Before 'pullos' 'secum' deleted.

*Lon iski,* see above, equal in size to the quail. The belly is white; the back is black adorned with blue spots, it has a longish body and it frequents lakes and rivers. It has a short tail and it dives in rivers to fish. See birds who frequent river banks in (Anglus Bartholomeus)[750]

Buinean liana ὀνοκρόταλος (the bittern).

The bittern is almost the same size as the guinea fowl in its type of plumage, but lean and unpleasant to eat. It is ash-coloured with long beak and legs. It frequents rivers and lakes and lives on little fish with which, when ejected from its first stomach, it rears its chicks. See Calepinus and Pliny.

*Creachoig: in Spanish, arandato, adorned with white and black and very scolding. See Aristotle on jackdaws smaller than the jackdaw (graculus), larger than the magpie. λύκος[752] see satiria.*

The pheasant[751] is a little smaller than the guinea hen, black in colour, distinguished by blue spots, it has eyelids clothed in red bristles, with a long beak and legs, it flies with great impetus by flapping its wings strongly. It has two types of flesh, the outer is black in colour, the inner white. There are people who consider it the most tasty and pleasing of all birds to eat. It has whitish eggs, distinguished by grey spots. It lays among the grasses and branches on the ground and the nest is not hidden in any way. Leading the chicks, while featherless, it feeds them and keeps them warm with its wings, it feeds on fruits, grains of flax, acorns, berries, hazel nuts, and other fruits. It often hatches out twelve chicks, sometimes less. It perches on trees.

*Grain catha, a little bird, dark in colour, a song bird perhaps a nightingale. Neaska (snipe) has a long beak and is dark in colour. See friguilla. L. Fringillago[753] manticula.[754] See also annotation. Gabhar roth, gabar deora (male snipe, gabhairín reodha), minnan eair, see caprimulgus: a goat-milker: Pl. 10, 40, 56 §115. The lagopus (redgrouse) in Ireland; it is of the family of partridge. Pl. 10,48,68. §123. There are birds of Diomedes in Ireland (see Aen. 11,271) [some of the companions of Diomedes were changed*

---

750. Ken Nicholls suggests that this may be a diver.
751. The pheasant was previously described on page 53.
752. A kind of daw (Liddle & Scott).
753. Frinfilla, fringuilla or frigilla, robin redbreast or chaffinch.
754. Manticula is a small purse.

*Platea eius nomen quaere. Eadem forme pelicanus et onocratulus.*
*Cornix, frugivora finnin fearr, ver finnin feoir an dadrian eainin.*

Hispanice[755] mosquitillo minor passere, capite dorso, et alis fuscis, ventre, gutture, et mento albidis, digitis quaternis, cruribus mediocribus, rostro brevi.[756]

*Runc fiodhrinm giodhrinn ex genere marinorum anatum et anati par velea maior*

Haec avicula est revera luch liath, et luch cabliath, et ea minor, et eiusdem coloris est cno chuill, de qua infra. an eadem Hispe., moscareta vide Batis et rubetra.

*Gealbhon coiri gealbhon skibreach passer sub ruber.*
*Magnitudine passeres,*
*Kearc iski maoir conturnice colore, nigris rostro et pedibus brevibus, flumina brolit natat est maoir ĝ? lonisk, vide tringa*

Aetate nostra in Ibernia cum pueri onocrotali nidum, qui in uligine positus erat, invenissent, ova sex, quae in eo repererunt, abstulerunt: sed a maioribus natu iussi rursus in nidum quinque retulerunt e quibus avis totidem pullos exclusit. Hi abducti domi, pane, caseo, cocta carne et aliis cibariis pascebantur. Ova sunt cyanei coloris.

Cno chuill, avicula canora, regulo[757] maior,[758] eique colore fusco similis.

---

755. 'Luch liath and luch cibhliath' deleted before 'Hispanice'.
756. After 'brevi' 'est hic forte cno chuill de qua infra' is deleted.
757. Illegible word deleted after 'regulo', inserted above the line.
758. 'Colore' deleted after 'maoir'.

*into seabirds who haunted the Diomede islands off the Apulian promontory of Garganus. Diomedes, a hero of Troy, was punished as he wounded Aphrodite in the arm. See Iliad 5,318.]*[759]

*Platea: the same as pelican and onocratalus*

The flycatcher *mosquitillo* in Spanish, is smaller than the sparrow; the back, the head and wings are dark, the belly and throat and chin are whitish in colour, it has four toes, medium length legs and a short beak.

This little bird is in fact the *luch liath* and the *luch cabliath* (unknown bird) but smaller than it and the same colour as a *cno chuill* (unknown bird). See below, and whether it is the same as the Spanish, *moscoreta* (flycatcher). See, the *batis* (stonechat) and *rubetra*.

*Cornix frugivora (rook)*
*Finnin feor (carrion crow)*
*Finnin feair (Carrion crow)*
*Dadrian eainin (dhá drian éainín rook)*
*Fiodhrinn or giodhrinn (giughrann, barnacle goose) from the family of sea ducks equal to or bigger than a duck.*

*Gealbhan coilli (hedge sparrow)*
*Gealbhan skiebrach (gealbhan breac, brambling euarasian finch)*

In our time, in Ireland, when boys found the nest of a pelican which was placed on moist ground, the six eggs which they had found there, they took away, but on the orders of their parents they put five back in the nest from which the bird hatched out all the chicks. When these were taken home, they were fed with bread, cheese, cooked meat and other bits of food. The eggs are dark blue in colour.

*Passer subruber (a reddish sparrow), the size of a house-sparrow*
*Coislean, coiri, seacon, field-fare, kearch iski.*
The waterhen is larger than the quail. Black in colour with short beak and legs. It inhabits rivers and swims; it is larger than the water ouzel (dipper). *Lon iski.* See *tringa.*
*Keán choil crossed out under Keán choil.*[760]

---

759.   This section is crossed out in the manuscript.
760.   Ken Nicholls suggests that this may be the green woodpecker.

761

(Kean choil magnitudine meruli pectore flavo)[762]
circum palpebras circuitu flavo summo vertice rubro, similis alis
rubris, caeteris fuscis citima cauda flava, rostro fusco, pedibus
albis unguibus longis, arbores tundit, in earum cavis nidificat,
ex ramis sese pedibus suspendit, multos iaculatorum iactus
expectat. Est macra. Vide Galgulus.)
*Deark gdharraig avicula magnitudine par sylvae et regulo,*[763] *colore
fusco, dorsi ventris albido figit ungulas arboribus balga re crann vidi
parus et Hispanice chaidariz.*

*Lacha cruoigh*[764]
*chraonn vide*[765]

Coilchenn columbae magnitudine, plumosa et macra, variis
plumarum coloribus, albo, caeruleo, purpureo, aureo, et
aliis ornata, avis pulcherrima. Sunt qui eam omnibus avibus
pulchritudine praestare autument. Vide Serkin do chum dia, et
fear coimedi na nian

*Livia. VinagoColman,
vide Aristotelum et
dict. Anglum Caog
Airni, Hisp. Canaria,
Ib Gioboglin, Boiglin
Anetiam, Buioc Cinn
Dir.*

Duibhian lacus frequentat, avis macra, et esu ingrata
magnitudine anseris, colore nigro omnino. Vide plalacrocarox.

From the manuscript: Agunt de Purgatorio divi Patricii.[766]
'Divus Antoninus Arch Episcopus Florentinus ordinis
praedicatorum in summar theological tomo 4. titulo 14 c. 10
8. Vincentius burgundus praesul belua censis ordinis prae
dicatorum in speculo maiore tomo 4 lib 20 c. 24, Jacobo
Januensis ordinis praedicatorum in legendis sanctorum c. 49 in
vita d Patricii ubi refort Nicolai qui purgatorum fuit ingressus.'

---

761. The words in brackets are inserted from the L. margin.
762. 'Kean choil' is deleted after kean choi1.1
763. 'Minor fere' deleted before 'regulo'.
764. An illegible word and lacha chroin are deleted and lacha chruigh
      inserted above the line.
765. Two illegible words are deleted after 'vide'.
766. This section on the purgatory of St. Patrick is inserted here in the
      MS but obviously should be inserted before C. Liv Purgatorium Divi
      Patricii. So I have moved it there as well as keeping it in its original
      position in the MS, which is out of place among the birds. The author
      may have left blank spaces which were filled in by somebody with
      a different hand and so the incongruous description of birds. The
      pagination is in correct order.

*Keán choil*

The keann choil (ceann caol) is the size of a blackbird with a yellow breast, round the eyelids there is a yellow circle, the top of his head and the tips of the wings are red, the rest dark, the tail nearest the body is yellow, white legs and long nails. It hammers on trees and nests in their cavities. It suspends itself from the branches by its legs. It awaits the many attacks of fowlers. It is lean. See Galgulus.[767]

*The tom tit is a little woodland bird equal in size to the robin and the wren, dark in colour, white on the back and belly. It fixes its claws on trees. The tree creeper see parus (tomtit) in Spanish chamariz.*

The *cnó chuill* (unknown bird), a little song bird, larger than the wren and similar to it in a dark colour.

The yellow hammer[768] is the size of a blackbird, except that it is more graceful and longer; it has a golden breast. See *Galgulus* and *icteris*.

(Coilchenn, cailicín), the sea pigeon, is the size of a dove; it has fine feathers and is lean, adorned with varying colours of its plumage, white, blue, purple, gold and others; it is a very beautiful bird. There are people who say that it outshines all birds in beauty. See Serkin, created by God, and the protector of the birds.

*Widgeon (laca chruaigh) Teal (chraonn lacha) Livia vinago, colman dove*

Duibhian (dubh éan), the cormorant frequents lakes, it is a lean bird and unpleasant to eat, it is the size of a goose, completely black in colour. See Phalacrocorax (coot or cormorant).[769]

*See Aristotle said Bartholomaeus Anglus. Crow with a red beak (cnogairm pyrrlocarax – chough.)*

St. Antonius, the Archbishop of Florence of the order of preachers in the Summa Theologica in Vol. 4, Titulo 14, Chapter 10 ? 8, Vincent of Burgundy Bishop of Beauvais of the order of preachers in the Legenda Sanctorum C. 49 in the life of St. Patrick where he recounts the story of Nicolaus who went into purgatory. (I have transfered this to p. 249 before 'The Purgatory of St. Patrick' where I felt Don Philip intended it to be. He may have left some empty pages intending to expand further on Nicolaus and purgatory here but somebody in a different hand returned to a description of birds.)

*In Spanish canario (canary) Irish gioboglin, bioglin; perhaps also yellow hammer, búiog an cinn óir. Agunt de Purgatorio Divi Patricii*

---

767. Galculus, previously described on p. 57 marginal, again O'Sullivan Beare confuses the Galgulus (witwall) with the i)/kteroj (golden oriole). See Pl. 20,11,28 and Mart. 13,68.

768. Buibhian (Buidhéan) is the word Don Philip uses here and that means a yellowhammer. Ken Nicholls thinks, on the basis of the description, that the bird may be the golden oriole.

769. Pliny. book 10. 48, 68; book 11. 37, 47.

(Agunt de Purgatorio divi Patricii Divus Antonius, IE L. 35, c.8, ?.27,, L.19, c.2. Divus Dionysius carthusianus Liq de Novissimis Art. 48, e de indicio animae art. 24. Aquatinus doctur antiquus Lib. 3, c.240, Frater dira Franciscanus de Purgatorio c.26, Pater Sanches Jesuma, Folio 454, Numero 61 et Folio 617.)

*Riobog nonest*
*calandra sed avicula*
*eiusdem fere coloris*
*passere minor et*
*gracili corpusculo*

Calandria cassita maior ventre albido dorso fusco nigris punctis distincto, longula nigrioreque cauda, rostro fusco brevi, cruribus mediocribus, quaternis digitis, optime canit. Vocatur Ibernice smolach et Hispanice calantria, quare nomen latinum. Vide χάλανδρα (alauda et κορδαλός)

*Grain choirrh*[770]

Pardillo est corpusculo longiore, quam passer, toto colore fusco, cauda longiuscula, rostro brevi canit interdium vereque melius, nocte silet, digitis quaternis.

*Glasain cuilinn*
*an?*[771] *habitat in*
*sylvis et sepibus*

Verdecillo, gafarron est breviore corpusculo, quam pardillo. Eius color ex viridi et fusco, vix micante flavo, componitur, longula cauda, bene canit.

*est gracilior minor*
*passere*

(Vide vireo Coinllean, crosach et caochan giomhbas magnitudine sylviae colore ex viridi et fusco.
Hoc et etiam regulo minor est coinllean catha.[772] Miuntan oir uterque, ille maior hic minor.

*Coinnleor muire,*
*finiin oir an*

Chlochran chlochan chloch[773] punctis varius, cauda[774] citima alba, ultima nigra, ventre albido alis et dorso viridibus albidis maculis distinctis magnitudine alaude).[775]

---

770.   'Luch lieth and' deleted before 'grain choirrh'.
771.   Deleted after 'an' 'crosac, coillan catha'
772.   'Vide? aegidus' deleted after 'coinllean catha'.
773.   'Viridus altudis?' deleted after 'chlochan cloch'.
774.   '?Intima nigra extima' deleted after 'cauda' and 'citima' inserted above the line.
775.   From 'Vide vireo' down to 'magnitudine alaude' in brackets is inserted from the R. margin.

*Riobog (riobbog) the*
*hedge sparrow is not*
*the lark but a little*
*bird similar to it in*
*colour, smaller than*
*a sparrow with a*
*slender body.*

The lark, larger than the tufted lark, with a whitish belly and dark back, distinguished by black spots with a longish and blacker tail with a short dark beak, medium length legs, four toes and sings excellently. It is called, in Ireland, the *smolach* (thrush),[776] in Spanish *calantria;* look up the Latin name. See χαλανδρα (woodlark), *alauda* (crested lark) and κορνδαλο/φ (crested lark).

*Gráin choirrh*

The *pardillo* (the linnet) has a small body, longer than the sparrow, completely dark in colour a slightly long tail, a short beak. He sings by day, rather sweetly indeed, he is silent at night, he has four toes.

*The glasain cuillinn?*
*It lives in woods and*
*hedgerows. It is more*
*slender, smaller than*
*the sparrow.*

*Verdecillo,*[777] *gaffaron*[778] has a small body, shorter than the pardillo. Its colour is composed of green and dark with a glint of yellow. It has a longish tail and sings well. See *vireo* (green finch), *coinllean crosach* and caochan, *giombhas* (green finch) the size of a *sylvia*[779] in colour between green and dark.

The *coinllean catha* (coinnleán cátha, coinnlinn cátha), yellow hammer, is even smaller than the wren. Each is called *miuntan,* the former is larger, the latter smaller.

*Clochran chloch, clocháncloch,* the wheatear, varicoloured, spotted. The tail is white at the base and black at the tip, whitish belly, wings and back marked with the greenish whitish spots. In size, it is equal to the crested lark (*sic*).

---

776.   Obviously wrong, as *smólach* is a thrush.
777.   Could not find it.
778.   Could not find it.
779.   *Sylvia* according to O'Sullivan Beare is a robin see p. 55 (under Small Birds ch. 22); in modern terminology it refers to the warblers.

Gilguero verdecillo magnitudine par longula cauda albido
ventre: alis nigro et aureo coloria variatis: fronte, et mento
rubro, nigro distinguente. Canora avicula.

The *gilguero*, equal in size to the *verdecillo*, it has a longish tail with a whitish belly: it has wings variegated with black and gold. Its forehead and chin are red picked out with black. It is a small song bird.

## C. LIII
### Miracula

Ex uberiore Iberniae supellectile rerum mirabilium haec recensui, quae quidem quamvis laudemus, non tamen stupemus, quia solito naturae cursu genita conspiciuntur. Habet autem illa insula alia inusitata, rara, occultiores causas habentia: ob id afferentia stuporem, indeque miracula nominata, quod admiratione digna sint. Quorum ego nonnulla alio opere divulgavi; sed nunc plura, etsi non omnia, repetere duxi.

### Agunt de Purgatorio divi Patricii[780]

Divus Antoninus Archiepiscopus Florentinus ordinis praedicatorum in summa theologica tomo 4. titulo 14.c.10 8 Vincentius Burgundus praesul Beluacensis ordinis Praedicatorum in Speculo maiore tomo 4. lib. 20.c.24 Jacobus Januensis ordinis Praedicatorum in legendis sanctorum c.49 in vita D. Patritii ubi refert historiam Nicolai, qui Purgatorium fuit ingressus.

---

780.   Deleted in the L.margin: Agunt de Purgatorio divi Patricii divi Antonius, IE L. 35, c.8, s.27, L.19, c.2. Divus Dionysius Carthusianus Liq de Novissimis Art. 48, e de indicio animae art. 24. Aquitius doctur antiquus Lib. 3, c.240, Frater dira Franciscanus de Purgatorio c.52, Pater Sanches Jesuma, Folio 454, Numero 61 et Folio 617. I have brought this forward from p. 43, which is where it is in the original MS as this seems a more appropriate place for it to be.

**St Patrick's Purgatory discussed by the following.**

## C. LIII
### Miracles

From Ireland's more abundant store of wondrous things I have chosen the following. Although indeed we praise them nevertheless we are not amazed because they are seen to have been produced in the usual course of nature. However, that island has other unusual, rare things which have more hidden causes: because of this they bring on wonder and from that they are called miracles, because they are worthy of admiration. Of these things, I have published some elsewhere, but now I have decided to recollect more but still not all.

### Agunt de Purgatorio divi Patricii

St Antonius, the Archbishop of Florence, of the Order of Preachers, in the summa theologica in Volume 4, titulo 14, chapter 10, §8. Vincent of Burgundy, bishop of Beauvais of the Order of Preachers in the Speculum Maius in Volume 4, 20, 24. James of Genoa of the Order of Preachers in the Legenda Sanctorum c.49 in the life of St Patrick where he recounts the story of Nicolaus who went into The Purgatory.

# C. LIV
## Purgatorium divi Patritii

*Purgatorium divi*      Itaque quod antea loco ultimo constitui, nunc primo colloco
*Patritii*     Purgatorium divi Patritii.[781]

*Lacus Deargus*     (Est in Ultonia satis noto Iberniae regno lacus nomine Deargus
*Loch Dearg*     forma pene circularis mille circiter passibus longus in eo lacu
est insula templo, divique Abeogi sepulchro venerabilis: Iuxta
quam est alia minor sita. In hanc Patritius[782] divinitus delatus, et
iussus)[783] locum prius baculo circumscriptum fodere et humum
eggerere institit, donec succedentem specum offenderit: et eam
qui pecatis a sacerdote expiatus rite ingrederetur, hunc Tartari
cruciamenta visurum, et si ea forti, fidoque anime sustineret,
commissorum piaculorum veniam impetraturum fuisse,

Specum vallo, et foribus instruxit. Iuxta eam templum
conditum Canonicis regularibus habitandum commisit, et
antri claves tradidit. In illud priusquam Patritius in coelitum
coetum migravit, non nulli descenderunt: quorum alteri fide
fluxa, et imbecilla amplius non extiterant; alteri christiano,
firmoque animo praediti,[784] orci calamitates, quas viderant,
reversi crebro referebant. Quod vero scrobis huius descensus

---

781.     Deleted here after Divi Patricii and difficult to decipher. 'Hic Iberniae
gentis clarissimus atque sanctissimus praeses, a Caelestino primi
pontifice summo christiani nominis illustrandi causa sub annum
redemptoris quadringentesimum missus in Iberniam ethnicos
homines ad Evangelicam veritatem illustrabat, vertere, conabatur,
cum alil referendo, tum Empyriae regiae summam faelicitatem,
et inferni calamitatem, poenas cruciatusque memorando. Quibus
rebus auditores aliqui dubii perplexique a divo petiverunt, ut averni
tormenta, quae tantopere verbis exaggerabat, ostenderet. Ille stolida
hominum petitione, maestus, Numinis fidem implorat, intentissimas
preces fundendo, corpusque totum nudum, durisque vereberibus
immolando. Neque frustra. Nam a Deo optimo maximo per Deargum
lacum forma pene circularem mille circiter passibus longum in Ultonia
situm in Insulam defertur'.
782.     After 'Patritius' 'a deo uptimo maximo' is deleted and 'divinitus' is
inserted above it.
783.     The words in brackets are inserted from the L. margin.
784.     After 'praediti' reversi is deleted.

# C. LIV
## The Purgatory of St Patrick

*urgatory of St*
*atrick*

Thus, now I allot the purgatory of blessed Patrick the first place, which previously I put in the last place. (Deleted paragraph insert here, p. 64 MS)

*ough Derg*

There is in Ulster a well-known kingdom of Ireland a lake by the name of Derg. Its shape is almost circular about one mile across. In that lake there is an island with a church and a sepulchre of Venerable St Abeogus, next to which there is another smaller island situated. Into this, Patrick, under divine influence had been brought and previously ordered to dig out a place previously marked around by a stick and he devoted himself to carrying out the earth until he struck a cave beneath it; and he knew that whoever went into that cave, whose sins were duly forgiven by the priest, this man would see the tortures of hell and if he endured them with a brave and faithful spirit; he would receive pardon for the sins he had committed. He fitted the cave with a palisade and doors. He ordered a church to be built next to it to be the abode of the Canons Regular and handed over the keys of the cave. Before Patrick crossed over to the heavenly assembly, some people went down into the cave; of them, some, being of wavering and weak faith, did not survive further. Others endowed with Christian and firm spirit, when they returned frequently recounted the misfortunes of hell which they saw. Because the descent to this grave was most dangerous, the trench itself afterwards was blocked up and another established in which men might spend the whole night in less danger. Of this purgatory, writers not at all obscure have made mention before me; they have committed to writing the experiences of some who visited it. Quite widespread is the story of a monk who having been twisted with many tortures in this purgatory and afflicted with wounds, impressed on his mind and memory an image of these events so that, while he lived, he turned them over in his mind just as if he were looking

erat periculosissimus,[785] ipsa postea fuit obstructa et altera condita, in qua[786] minore periculo homines pernoctarent. Huius purgatorii ante me scriptores nihil obscuri meminerunt: et casus aliquorum, qui illud viserunt, litteris commendarunt. Satis est vulgata historia monachi, qui in hoc purgatorio multis cruciatibus tortus, atque vulneribus affectus earum rerum effigiem ita memoriae, et animo impressit, ut dum vixit, non aliter eas cogitatione evolveret, quam si praesentes oculis semper intueretur. Ita socios suos monachos ludo nimis deditos, vel gaudio exultantes videbat, vociferabatur, eos stolidos, et inscios esse, et si scirent, levia etiam peccata gravibus suppliciis puniri, nunquam fuisse lusuros, laetaturos, sive risuros.[787] (Egneus miles labores adivit)[788] quos hic[789] Dionysius Carthusianus memoriae prodidit. Ramonis Vicecomitis Hispani eventa ego in Compendio narravi Nunc alius aerumnas[790] ex Ibernico libro vitarum sanctorum, qui OSullevani Bearrae principis iussu perscriptus est, erutas in lucem emittam.

*Nicolai historiam* | Igitur longo post obitum divi Patritii tempore Nicolaus vir
*refert Jacobus* nobilis, qui multa peccata commiserat, pro eis eam poenam
*Januensis in vita D.* suscipere constituit, ut hoc Purgatorium adiret. Ad id corpus
*Patricii in legendis* suum octo dierum ieiunio, sicuti de more fieri solet, castigat. Inde
*sanctorum* aperto ostio in specum descendens, monachos albis vestibus indutos convenit. Hi postquam Deum laudibus extulerunt, Nicolaum animo forti, fideque constante esse iubent, illum cacademonibus saepe, et acriter quidem impetendum; semper tamen evasurum victorem, modo in summo discrimine positus Dei opem imploret, haec verba repetendo: *Domine Jesu Christe fili Dei miserere mihi peccatori:* Hominem vix monachi reliquerunt, cum horribili agmine diaboli circumstantes, salutant: et erga eum singularem amicitiam, et amorem praeseferentes qua venerat, redire iubent: contra serpentes, atque reliquas bestias, quae ibi erant, praesidium, seque deducturos illum incolumen promittunt. Quod ille consilium capere intrepide noluit. Tum manes ingentem clamorem, moestissimasque voces efferendo,

---

785.   After 'periculosissimus'.
786.   After 'in qua'.
787.   After 'risuros' 'egnei militis casus laboresque' are deleted.
788.   The words in brackets are inserted from the R. margin.
789.   After 'quos hic' 'adivit' plus an illegible word are deleted.
790.   After 'aerumnas' 'antehac, quod ego sciam, minime devulgatas' are deleted.

at them constantly present before his eyes. Thus, when he used to see his own companion monks given to too much play or exulting in joy, he used to shout that they were foolish and ignorant and if they should know that even venial sins were punished with severe penalties they would never play games, or be happy or laugh.

Egneus, a knight, underwent toils here which Denis the Carthusian has recorded. The experiences of Ramon the Spanish Viscount, I have narrated in the *Compendium*. Now I shall publish the hardships of another man that are found in an Irish book of the lives of the saints, which was written down on the order of the chieftain O'Sullivan Beare.

*Jacobus Jansenius records the story of Nicolaus in the Life of St Patrick in the Legenda Sanctorum.*

Therefore, a long time after the death of St Patrick, Nicolaus, a noble man who had committed many sins, decided to undergo the penance that he should visit this purgatory. To this end, for eight days, fasting, he punished his body, just as is the accepted custom. Then, when the gate was open, he descended into the cave and met monks clothed in white garments. Then, when they exalted God with praises, they ordered Nicolaus to be of strong spirit and firm faith: that he would have to fight with evil spirits often and sharply. However, that he would always turn out the victor. Even when placed in the greatest peril, he should implore the help of God by repeating the following words: 'Lord Jesus Christ, son of God, pity me a sinner'. The monks had scarcely left the man when the devils standing around in a frightening line saluted him, and, pretending outstanding friendship and love towards him, they ordered him to return home whence he had come, offering protection against the serpents and the rest of the beasts that were there they promised that they themselves would lead him back safe and sound. This plan,

Nicolao tantum terroris incusserunt, ut pavidus, totaque mente et omnibus artubus contremiscens humi ceciderit. Caeterum documenti, quod monachi dederant, memor verba ab eis praescripta sublata voce proferendo, non modo sese rursus levare, et erigere, sed etiam hostem in fugam vertere potuit.

Hinc in secundum locum venit: ubi statim insidiis hostium capitur, qui illi mortem quam crudelissimam intentant, nisi ipsis obtemperet. Ille minas contempsit. Hi in diffusissimum, atque patentissimum campum ardente igne oppletum[791] Nicolaum coniiciunt. Ille vociferans salutifera verba pronunciando, et ignem extinguit, et hostes fugere cogit.

*Virg. Lib. 6*

*Alium locum petit*
*Hinc exaudiri gemitus, et saeva sonare*
*Verbera: tum stridor ferri, tractaeque catenae*
Hic erant homines quamplurimi vincti hortibles visu formae. Hi ex flammis in flammas iaciuntur; illi ventrem cum dorso coniunctum habentes, eisdem continenter ignibus comburuntur, praeque doloris magnitudine humum mordent. Alii ferreis verberibus caesi singulis ictibus usque ad viscera scinduntur. Alii a serpentibus rosi comeduntur. Has poenas Eumenides Nicholao minantur, nisi inceptis desistat: renuentemque in medias flammas inter bestias iaculantur. Ille pristinis verbis prolatus incolumis evadans, hostes dare terga compellit.[792]

---

791. After 'oppletum' 'in?' deleted.
792. Deleted after compellit: In alium locum delatus profundum horrendissimumque puteum, qui *sartago* vocabatur, flammas vomentem, in eo alios genuum tenus, alios umbilico tenus multos ad pectora usque, non paucos usque ad mendum, quosdam etiam usque ad tempora mersos conspicit. Hunc Belsebubo tartareorum principi domicilium addictum esse: ex eo neminem redimi, vel regredi, in illum se Nicolaum, nis ipsis morem gerat, dimissuros caco daemones asseverant, & obsequi renuentem detrudunt. Caeterum cum Nicolaus conceptus verba repetijsset, inferi spiritus, atque cruciatus, evanuerunt.

bravely, he refused to adopt. Then, the spirits by letting out a great shout and the saddest voices struck so much terror into Nicolaus that, terrified and quaking, shaking violently, in his whole mind and in all his limbs, he fell on the ground. Besides, mindful of the teaching which the monks had given, in a loud voice giving out the word prescribed by them, not only was he able to get up again and straighten himself but even to turn the enemy to flight.

From there he came to the second place, where straight away he is taken by an ambush of the enemy who threaten a most cruel death for him unless he would submit to them. He showed contempt for the threats. They threw Nicholaus into a very wide and very open space filled with burning fire. He, shouting and pronouncing the health-bringing words, both extinguished the fire and compelled the enemies to flee.

*Vir. book 6, 557–58* *He seeks another place. From there, groans were clearly heard. There was the sound of savage blows. Then, the screech of iron and dragging of chains.*

Here were very many men bound, shapes horrible to behold. Some were being thrown from flames to flames. Others, tied belly to back, were continuously being burned with the same fires and because of the greatness of their suffering they were biting the soil. Others, lashed with cruel blows, were split right to the guts with each stroke. Others, having been gnawed by serpents were being devoured. The furies were threatening Nicholaus with these penalties unless he stopped what he had undertaken, and as he refused, they were throwing him into the middle of the flames among the beasts. He, borne forward by the same words, escaping safe and sound, compelled the enemy to show their backs.

Inde pervenit ad pontem nimis angustum, et non minus, quam gelu, lubricum, sub quo sulphurei ignis foetidus amnis fluebat. Eum cum iuberetur transire, terrore perculsus stupet, mentis incompos, et divinorum verborum immemor. Nihilominus aggreditur. Hominis pondere pons partim deprimitur, partim tollitur, totus tremet. Quae pars incedentis pedibus calcatur, mox illa gemit, stridet, rumpitur. Caeterum Nicolaus linquentem prae pavore spiritum revocat, sacrorumque verborum recordatur: quae quoties profert, toties pars pontis quam terit, reficitur, redintegratur, et firmior quiescit. Ita Dei ope pontem traiicit.

Unde progressus pervenit in campum pulcherrimum, magnaque fragrantia laetum.
*Devenere locos laetos, et amoena vireta*
*Fortunatorum nemorum, sedesque beatas*
*Largior hic campos aether et lumine vestit*
*Purpureo: solemque suum sua sydera norunt*

Hic duo iuvenes lucidissimo candore fulgentes obviam facti eum ad[793] regiam[794] auratis moenibus cinctam ducunt. E regiae porta iucundissimum odorem emissum cum Nicolaus hausisset, mox ab omnibus praeteritis laboribus, terroribus, atque cruciatibus sese omnino integrum, et refectum sensit. Intrare vero volenti, iuvenes[795] dixerunt, illum esse paradisum, introireque minime licere, donec prius apud mortales[796] triginta dies agat; et rediro iam[797] integrum esse, nullibique spiritus hostiles nocituros, sed eius potius conspectum fugituros: Nicolaus eadem, qua iverat, daemoniis sese passim fugae mandantibus, nemineque prohibente, in primum[798] specum revertitur:[799] et ad mortales

---

793.  'Ad' is inserted above the line.
794.  'Versis' deleted after regiam.
795.  'Qui angeli erant' deleted after iuvenes.
796.  Mortales inserted above the line.
797.  'Iam' is deleted in the line and reinserted above the line. 'Illi' which was inserted after 'iam' above the line is also deleted.
798.  'Im primum' is inserted above the line.
799.  The 're' of revertitur is inserted above the line.

From there he came to a very narrow bridge which was as slippery as ice under which a foul river of sulphuric fire was flowing. When he was ordered to cross over, he was astounded, struck by terror: he lost his mind and was unmindful of the sacred words. Nevertheless, he goes forward. The bridge was partly pressed down by the weight of the man, partly it held up – the whole thing trembled. The part which was pressed down by the footsteps of the man as he proceeded, soon that part groaned, creaked and burst. But, Nicolaus recalled his spirit, which was softening through fear, and remembered the sacred words. As often as he produced these words so often the part of the bridge that he trod on was renewed, put back together, steadied and grew stronger. Thus, he crossed the bridge with the help of God.

*Virg. VI, 638–41*

Having gone on from there he came into a most beautiful plain, pleasant, with a great fragrance.
'And now they arrived at the land of joy,
the pleasant green places of the fortunate woods,
the homes of the blest. Here, a more bountiful air
clothes the plains in a brilliant light and they are
aware of a sun and stars that are theirs alone.'

Here, two young men shining with a clear dazzling whiteness stood in front of him and led him to a palace surrounded by golden walls. When Nicolaus drew in the pleasant odour from the doors of the palace, soon he felt himself completely sound and restored from all his past labours, terrors and tortures. The young men said that indeed there was entrance to one who truly wishes, that that was paradise but no permission was given to enter until first of all he spend thirty days among mortal men; but return was now safe, and nowhere would the hostile spirits harm him but would rather flee at the sight of him. Nicolaus returned the way he had gone, with the demons themselves everywhere taking flight, and since no one prevented him, he came back to the first cave and when he came back to mortal

regressus, quae vidit, saepe enarravit, affirmans, se intra trigesimum diem mortem obiturum. Quae dies cum venisset mundo, diaboloque victis ab hac vita[800] migravit.

## C. LV
### In Ibernia nullum animal venenosum

*Venenosa animalia*
*nulla*

Est illa totius insulae laus insignis, quae venenosum animal nullum vel generat, vel importatum sustinet vivum, quin et suo pulvere, et rebus aliis, quas profert, peregre missis virosa extinguit, ut alio opere fusius a nobis est explicatum. Quam mirificam atque salutarem vim ipsius insulae naturae initio inditam fuisse, iste Gyraldus falso putat: cum postea divi Patritii meritis a Deo concessam esse liqueat.

*Ibernici libri et*
*Jocelinus*

Patritius enim[801] constante veterum Ibernicorum monumentorum fide baculo (qui, quod a Christo Domino sit ad[802] illum datus, Jesu dicitur) sublato[803] Dei ope, et angelorum auxilio omina Iberniae venenosa viventia in editum montem mari immanentem, qui Ibernice tunc *Cruchan Oeghli,* hodie vero *Cruach Phadraig,* idest[804] mons Patritii nuncupatur, coegit: indeque obruenda in pelagus compulit. Eiusdem sanctissimi nostri praesidis ope, spero etiam venenosa animalia, quae non modo corpora; sed etiam animas haeresis viro feriunt, Anglos ex sacra insula brevi esse deiiciendos: qua de re carmen illud a mea conditum fuisse, recordor.
*Patritius colubros terrae delevit Iberniae*
*Corpora qui viro laedere morsa solent*
*Horrida lethiferos huc missitat Angliasaepes,*
*Qui mordent animas, interimuntque pias.*
*Hos etiam meritis pellendos praesulis almi,*
*Spero telluris finibus esse sacrae.*

---

800.   'Incolum' should be 'incolumis' is deleted and 'ab hac vita' is inserted.
801.   After 'enim' 'veteribus Iberniae' is deleted.
802.   'Ad' inserted above the line.
803.   After 'sublato' 'dei ope' is deleted and reinstated.
804.   Before 'idest' 'est' is deleted.

men, he often told what he saw, confirming that before the thirtieth day he would die. When this day came on earth, having overcome the world and the devil, he left this life.

## C. LV
### In Ireland, no poisonous animal

*No poisonous
animals*

There is that outstanding praise of the whole island which neither breeds any poisonous animal nor sustains one living, if imported; rather if its dust and other things which it produces are sent abroad, they are an antidote to poison as has been more widely explained by me in another work. That man, Gyraldus, falsely thought that such a wondrous and health-giving force has been instilled in the nature of the island itself from the beginning, since it is clear that it was given afterwards by God through the merits of Saint Patrick.

*Irish books and
Jocelinus*

For Patrick, according to the firm faith of the ancient Irish annals, having raised up his crozier with the help of God (a crozier which was called Jesus because it had been given to him by Christ the Lord) and with the help of the angels he drove all venomous living creatures of Ireland into the high mountain overhanging the sea, a mountain which in Irish at that time was called Croach Oeghli, today indeed Croagh Patrick, that is the mountain of Patrick. From there, he drove them into the sea to be destroyed. Through the help of the same, our most holy bishop, I hope that also the venomous animals which strike not only the bodies but also the souls with the poison of heresy, i.e. the English, will be thrown out from the sacred island shortly. For this reason I record a poem that was composed by me:
*Patrick destroyed the serpents of Ireland, who are wont to harm with poison the bodies they bite. Dreadful England sends death-bearing serpents here which gnaw and kill holy souls. However, I hope that they are driven out from the boundaries of the sacred land by the merits of our kind bishop.*

## C. LVI
### Loca mirabilia

*Mures ubi non?*

Mures, quos nos Ibernicorum maximos supra diximus, et vulgus Ibernorum Gallicos vocat, nullos vel Ardmacha urbs, vel agri illi circumiacentes ferunt: quod et in aliis Iberniae locis contingere, proditur.

(Loca, in quibus mortuorum corpora minime putrescunt, et alia in quibus mox in cineres vertuntur, Ibernia habet.)[805]

## C. LVII
### Insulae miribiles

*Vermes ubi non?*

In insula, quam in Deargo lacu divi Abeogi sepulchro, insignem esse monstravimus, est sanctorum hominum sepluchretum, quo nullus vermis, nulla bestiola immunda vitae impunitate ingreditur.

*Duae insulae altera viventium*

In Momoniis insulae duae lacu cinguntur, antiquitis sancta religione clarae.[806] Alteram nullum foeminei sexus animal audet adire. In altera nemo moritur (unde insula viventium nomen accepit). In ea vero moribundi tot dolorum cruciamentis torquentur, ut malint eam efflandae animae causa relinquere, quam vitam tam acerbam ducere.

---

805.   The words in brackets are inserted from the R. margin.
806.   Before 'clarae', 'clarae' is deleted.

## C. LVI
### Wonderful Places

*Mice, where are they not?*

Mice, the largest of the Irish ones which were discussed previously, and which the ordinary Irish people call rats, are produced neither by the city of Armagh nor the surrounding lands, a feature that is reported also to pertain in other areas of Ireland. In Ireland are found places in which the bodies of the dead do not rot and others in which soon they are turned into dust.

## C. LVII
### Wonderful Islands

*Worms, where are they not?*

In the island in Lough Derg, which we have shown to be famous for the tomb of Saint Abeogius, there is a graveyard of holy men into which no worm, no foul small animal goes with immunity from death (i.e. it cannot survive).

*Two islands: one of the living.*

In Munster there are two islands, surrounded by a lake, famous from long ago for holy religion. Into one, no animal of the female sex dares to enter. In the other, nobody dies (from which it gets the name the island of the living). In that island, those about to die are tortured with so much affliction of pain that they prefer to leave in order to die than to prolong such a bitter life.

Also, it is reported that neither the fish of this lake can be boiled, neither on the island of the living nor elsewhere in water from the same lake. In Connaught also it is reported that there is an island surrounded by a lake, and it is related that men cannot die until they are carried off it.

# C. LVIII
## Lacus mirabiles

In Ultonia Neachus est lacus pulcherrimus in quo(si Aquifoliae bacciferae rami, vel surculi inferantur, ex eis duodecimo mense)[807] arbores nasci tradunt summa parte aquis extante ligneas, infima terrae infixa lapideas, meda aquis cincta ferreas.

(Huis etiam lacus pisces nec in viventium insula nec eiusdem lacus aqua aliubi posse elixari feruntur. In Connacte quoque insulam lacu cingi proditum est e qua donec homines efferantur non mori traditur in alio ultoniae lacu pomariuma est cuius arbores dulcia poma bis quotannis ferunt.)
*(Hic variant scriptores quae pars lignea vel ferrea sit vide mauritium et Bartholomeum Anglicum).*[808]

Alium in Ibernia quidem meminit lacum esse, in quem iniectae virgae colurnae in fraxinease, et fraxineae in colurnas vertantur.

*Alius, lacus mirabilis*  In Bearrae principatus lacu (cui Mochellogi templum immi-
*lacus*          net),[809] tres cespites virides, atque juncosi per multa saecula natare solent, etiam adverso vento.

# C. LIX
## Fontes mirabiles

*Lacus*   Si est isti Gyraldo fides habenda in Connachta fons est liquore dulcis in celsi montis vertice erumpens, qui quotidie bis aquis defectus, et totius abundans, maris vicissitudines imitatur: et in
*Puteus*   Iberniae parte, quae Angliae Guallias proxime respicit, puteus,

---

807. The words in brackets are inserted from the R. margin.
808. From 'huis etiam' to 'bis quotannis ferunt' are inserted from the L. margin. From 'Hic variant' to 'Bartholomeum Anglicum' is inserted from the R. margin.
809. 'Cui Mochellogi templum imminet', is inserted from the L. margin.

# C. LVIII
## Wonderful Lakes

In Ulster, Loch Neagh is the most beautiful lake in which, if the branches of berry-bearing holly or a slip are brought in, from those after a year trees are reported to grow which are wooden in the upper part that stands out from the water, stone in the bottom part which is fixed to the earth and iron in the middle part surrounded by water.

*In another Lake in Ulster there is an orchard whose trees produce sweet apples twice yearly. Here writers disagree as to which part is made of wood and which part is made of iron. See Maurice O'Fihely and Bartholomeus Anglicus.*

One remembers another lake in Ireland into which when twigs of hazel wood are thrown, they are turned into ash and ash is turned into hazel wood.

*Another*
*extraordinary lake*

In a lake in the chieftainship of Beara which the church of Mochellogue overhangs, three green sods full of rushes were wont to float, through many ages, even though the wind was against them.

# C. LIX
## Wonderful Wells

*A lake*

If trust is to be put in that man, Gyraldus, there is in Connaught a well with fresh water bursting from the top of a high mountain which twice every day goes dry and fills up completely, imitating the changes of the tide. And in that part of Ireland which looks back, at its nearest point, to Wales, a well, which against the

qui communi natural ordine inverso aquas in maris refluxu recipit, et fluxu emittit.

*15 Fons*    Fons in Momoniis memoratur, cuius unda abluti cito canescunt. Quibus et duos in ultonia addunt: altero loti nulla canicie

*16 Fons*    inficiuntur: in altero duae tructae multa saecula vixerunt,[810] altera fuit ab Anglis occisa: altera hodie apparet: utramque[811]

*17 Fons*    alicui divo in deliciis fuisse, aiunt.

*18 Fons*    Caeterorum fontium, quorum aqua gelida epota, sparsaque aegri curantur, salubrem vim, patrocinio divorum, quibus dicati sunt, acceptam refero.

## C. LX
### Aves mirabiles

*19 Aves*    Inter prodigia recensendas puto durridinas aves, quae quemadmodum ex pineis[812] lignis mari dui fluctuantibus nascantur, supra docuimus.

*20 Insula*    In Skealaga insula Iberniae adiacente supra fanum, et eius vallum Michaeli archangelo dicatum nullae aves possunt volare; sed alarum usu destitutae sidunt, donec illud pedibus transeant. Cuiusque rei naturalem causam alii anxie quaerunt. Ego a senioribus accepi illud beneficium a Deo miraculose collatum esse, in virum sanctum, quem in ea insula divinarum rerum meditatione vacantem cibus defecit.

---

810.   After 'vixerunt' 'doner' is deleted.
811.   After 'utramque' 'apud aliquem divum in? diei' is deleted.
812.   '? Abiegnis' deleted and 'pineis' inserted above it.

usual rule of nature receives waters as the tide goes out and sends it away as the tide fills.

A fountain in Munster is recalled: people who wash in the water quickly go grey. To which they add two in Ulster: in one of which if people wash, their hair never goes grey. In the other, two trout lived for many ages: one was killed by the English, the other appears today; both, they say, were the favourites of some saint.

I report that the health-giving force of wells whose cold water, when drunk and sprinkled on, cures the sick, is received through the patronage of the saints to whom the wells are dedicated.

## C. LX
### Wonderful Birds

Among the prodigies, I consider that the birds called barnacle geese should be examined again. We have shown above how they are born from pine logs floating a long time on the sea.

In the Skellig Island, adjacent to Ireland, no birds are able to fly above the holy place and its rampart dedicated to Michael the Archangel; but they settle, destitute of the use of their wings, until they cross over it on foot. Others anxiously seek the natural cause of this event. I learned from my ancestors that this was a gift miraculously handed down from God to a holy man who, while sojourning on this island for the meditation of divine affairs, was lacking in food.

# C. LXI
## Mirabile sepulchrum

*21 Sepulcrum*

Ex divi Senani sepulchro, quod in Iniscathacha insula ad urbis Lomnachae portum est, qui lapillum gerat, hunc non esse aquis obruendum, modo de divina fide bene sentiat, creditur. Sed quid divorum miracula insertor? Ea hic propemodum innumera sunt, et in aliquorum[813] vitis a nobis, ut spero partim enarranda.

# C. LXII
## Huius libri epilogus

Haec ego, quae ad texendam Iberniae descriptionem de illius nominibus[814] fertilitate, temperatione,[815] situ, figura, regionibus, episcopatibus, oppidis, montibus, sylvis, portubus, lacubus, fluminibus, fontibus, animantibus terrestribus, volatilibus, insectis, aqautilibus, arboribus, frumentis, leguminibus, herbis, his, quae ex aqua gignuntur, metallis, terrae generibus, lapidibus, gemmis, atque miraculis in hunc locum congererem, habui. Quibus etsi non omnes[816] eius laudes sum complexus, at plura ornamenta, quam alius ullus (quod sciam) perstrinxi. Hic autem meus labor, quamquam[817] alicui minus perfectus erit existimatus, eum tamen[818] novarum rerum accessione augere, quam totum opus et incipere, et perficere, multo facilius fore, nemo[819] negabit. Ego quod suscepi, assecutus mihi videor, ut Iberniam non desertam, inviam, et aquosam, sicuti Gyraldus volebat, sed innumeris laudem titulis[820] accumulatam. quin et illi Angliam nec ubertate agri praeferendam, nec clementia coeli comparandam; infestatione vero venenosorum, et numero miraculorum postponendam esse, liquido ostenderem.

---

813. 'Ipsius' deleted and 'aliquorum' inserted above the line.
814. 'Nominibus' is inserted above the line.
815. After 'temperatione' 'nominibus' is deleted.
816. After 'omnes' 'illius laud' is deleted.
817. 'Quamquam alicui' is inserted above the line. An illegible word is deleted in the line.
818. 'Tamen' is inserted above the line.
819. Nemo is inserted above the line and nullus deleted on the line.
820. 'Innumeris laudem titulis' is inserted above the line and 'multiplici memorabilium rerum supellectile' is crossed out in the line.

# C. LXI
## A Wonderful Tomb

It is believed that whoever carries a pebble from the tomb of St Senan, in the island of Scattery, towards the port of the city of Limerick, cannot be drowned as long as he maintains divine faith. But why should I pursue the miracles of the saints? These things here are almost innumerable and in the lives of some, as I hope, they will be partly told by me.

# C. LXII
## The Epilogue of this Book

These things I had which I could gather here in order to weave together a description of Ireland concerning its names, its fertility, temperateness, situation, outline, regions, bishoprics, towns, mountains, woods, ports, lakes, rivers, springs, terrestrial and flying creatures, insects, aquatic beasts, trees, corn, leguminous plants, grasses, things born of water, metals, types of soil, stones, gems and wonders. Although I have not recorded all its praises but I have collected more ornaments than anybody else as far as I know. This my labour, however, although it will be considered less perfect by somebody else, nobody will deny that it will be easier to increase it by the addition of new facts than to begin and complete a whole work. What I have taken up I seem to have accomplished: that Ireland is not deserted, without roads, and boggy, as Gyraldus would have it, but that it is heaped with glory under many headings: and that I have shown clearly that England should not be preferred to it in the fertility of its soil nor compared to it in the clemency of its weather: and that England should be put behind Ireland because of the infestation of poisonous animals and the number of miracles. While I take pleasure in the commemoration of these events, the telling has grown to the size of a whole volume. For this reason, I decided that the

Quarum commemoratione rerum[821] dum delector, in integri voluminis magnitudinem oratio excessit. Quamobrem aliarum Gyraldi calumniarum confutationem in sequentes libres rei iciendam esse duxi.

---

821. After 'rerum' 'omnium' is deleted.

refutation of other calumnies of Gyraldus should be put back again to subsequent books.

# Authors Cited

**Aethicus**, also Ethicus Ister, wrote the *Cosmographia*. It professes to be an abridged translation of a Greek account of the wonderful travels of Ethicus or Aethicus Ister. The work was very popular in the ninth and tenth centuries and it probably had considerable influence on the geographical ideas of the time. He made use of matter found also in Solinus, Justin Orosius and Isidore of Seville. The date of composition is later than 630 because there is clear influence from the 'Etymlogies' of St Isidore of Seville.

**Bartholomeus Anglicus** was a thirteenth century encyclopaedist. He was a professor in Paris when he entered the newly-established Franciscan Order, later moving to Magdeburg. The date of his death is unknown. His most famous work, '*De Proprietaibus Rerum*' is an encyclopaedia of all the sciences of that time (c.1231). Theology, philosophy, medicine, astronomy, chronology, zoology, botany, geography, minerology are the subjects treated in the nineteen books of this work, drawing on the works of Jewish and Arabic authors as well as Greek and Roman.

**Petrus Apianus** (1495–1552) was also known as Peter Apian, Peter Bennewitz and Peter Bienewitz. He was a celebrated mathematician, astronomer and cosmographer. In 1524, Apian published his *Cosmographia seu Descriptio Totis orbis*, which was an introduction to astronomy, geography, cartography, surveying, navigation, weather and climate, the shape of the earth, map projections and mathematical instruments. The *Cosmographia* was republished by the Flemish cosmographer Gemma Frisius in 1533 in an expanded edition which became a best seller and was translated into all major European languages.

**The Venerable Bede** (672/3–735) was an historian and Doctor of the Church. His great work, the *Ecclesiastical history of the English People,* was completed in 731, and covers the period 597–731. It contains much that is relevant to Irish history, including the relationship of the Irish church to the papacy and to the church in England.

**Boethius, Anicius Manlius Severinus (480–524)** was a Roman statesman and philosopher often styled 'the last of the Romans', regarded by tradition as a Christian martyr. As early as 507 he was known as a learned man and as such was entrusted by King Theodoric with several important missions. However, when his enemies accused him of disloyalty he was cast into prison, condemned unheard, and executed on the orders of Theodoric. His treatise on *The Consolation of Philosophy* was extremely popular throughout the middle ages and early-modern period, as were his theological treatises, which include 'De Trinitate' and two short treatises (Opuscula) addressed to John the Deacon (afterwards Pope John I); Liber contra Eutychen et Nestorium and 'De Fide Catholica' (generally regarded as spurious, although the only argument against it's genuineness is lack of manuscript authority). He also contributed to the science of mathematics and the theory of music.

**George Buchanan (1506–82)** was a Scottish humanist educated at St Andrews and Paris. In 1536 he became tutor to James V's illegitimate son, James Stuart (later Earl of Murray). He was imprisoned (1539) for satirizing the Franciscans but escaped to the continent. He taught at Bordeaux where Montaigne was among his pupils and at Coimbra and became highly regarded as a Latin poet. Returning to Scotland, Buchanan declared himself a Protestant. He became an opponent of Mary Queen of Scots after the murder (1567) of Lord Darnley, and in 1571 he published *Detectio Mariae Reginae,* a bitter attack on the Queen. From 1570 to 1578 he was tutor to the young King James VI of Scotland, later King James S of England. His *Rerum Scoticarum Historia* (1582) is a useful source for history of his time and the source of O'Sullivan's quotation.

**Ambrogio Calepino (1440–1510/11)** was an Italian lexicographer and member of the Augustinian Order. His Latin dictionary, the *Cornucopiae,* appeared first in 1570 at Reggio. It was reprinted many times during the sixteenth century, and later editions were considerably enlarged. To the Latin editions were added equivalents in other languages. Calepinus also produced an edition of Pliny's natural history which was used by Don Philip, and wrote *The Life of St John the Hermit* which is found in the *Acta Sanctorum* for the 22nd of October.

**Giraldus Cambrensis,** born c.1146 in Manorbeer Pembrokeshire. His brothers and many of his cousins were prominent in the Norman invasion of Ireland. In his description of the invasion (Expugnatio Hiberniae) Giraldus always gives special praise to his own relatives. He was educated in Paris, and became a significant ecclesiastic on his return to Wales, though he was several times denied preferment to the bishopric of St David. In 1183 he paid his first visit to Ireland and he joined the entourage of Henry II in 1184, being employed partly in diplomatic negotiations with the Welsh and partly as tutor to Prince John with whom he came to Ireland for a second time in 1185. Under Richard I he was attached to the Bishop of Ely who administered the realm during the absence of the King in the Holy Land.

Giraldus wrote two books on Ireland, the *Topographia* and the *Expugnatio*. The *Topographia* is divided into three parts. The first part deals with the land of Ireland, its geographical position, climate, physical characteristics, flora, and fauna. The second part treats of wonders and miracles recorded in Ireland. The third part concerns the people of Ireland, their origins, customs, culture, and religious observances. The *Expugnatio* is an account of the conquest of Ireland from the Norman perspective. Although they were criticised in his own time, Giraldus' writings became popular in the sixteenth and seventeenth centuries owing to their dissemination in several humanist editions. They were valued because of the comparative lack of first-hand descriptions of Ireland, and perhaps because their criticisms of the Irish suited the interests of Elizabethan England. Many Irish writers wrote refutations, such as Peter Lombard, David Rothe, Thomas Messingham, Stephen White, John Lynch, and Geoffrey Keating.

**William Camden (1551–1623),** an English historian/antiquarian who was a founding member of the Society of Antiquaries. His major works are the *Britannia* and *Annales Rerum Anglicarum et Hibernicarum,* which resurrected the history of Ireland according to Giraldus Cambrensis. Although he added much useful detail, Camden also cast many aspersions against the Irish of his own day, and in this he was vigorously opposed by Geoffrey Keating.

**Edmund Campion (1540–1581)** was a precocious young scholar, educated at Oxford. He travelled to Ireland in 1569 where he stayed with James Stanihurst, the father of Richard Stanihurst who was a pupil of his at Oxford. Campion was distrusted as a papist and orders were used for his arrest but for a number of months he eluded pursuit. During this time he wrote a *History of Ireland* based on the work of Giraldus Cambrensis. It was written to prove that education was the only means of taming the Irish. This work was first printed by Richard

Stanihurst in *Holinshed's Chronicles* (1587) and then by Sir James Ware in his *History of Ireland* (1633).

In 1578 Campion was ordained Deacon and Priest by the Archbishop of Prague. Campion and another Jesuit, Parsons, were sent to England on a mission. He was captured in 1581 and executed after considerable torture. His work against the Protestant religion, the *Decem Rationes, et alia opuscula ejus selecta* was published at Antwerp (1631).

**Bartholo Cassineus, or Barthélemi de Chasseneaux, (1480–1541)** was President of the Parliment de Province. His early career was as the King's advocate at Atun, then he became Councellor of the Parliament of Paris and, finally, President of the Parliament de Provence until his death in 1541. His magnum opus was the *Catalogus Gloriae Mundi* from which Don Philip gets his quotations. This was first published in Lyon (1529).

**David Chambers (d. 1641).** A convert from Scots Presbyterianism, Chambers was ordained a priest in 1612 and undertook diplomatic work for the Holy See until 1623. Thereafter he was primarily associated with the Scots College in Paris where he was eventually appointed principal in 1637. He authored *De statu hominis veteris* (Chalon, 1627) and *De Scotorum fortitudine. pietate et doctrina* (Lyon, 1631). Chambers was a follower of Thomas Dempster whose *Scotorum scritporum nomenclatura* (Bologna, 1619) had already caused such a furious reaction from Irish writers. Though Dempster's work had since been placed on the Papal index, Chambers repeated his claims about *Scotia* and the ancient *Scoti*, seeking thereby to appropriate Ireland's saints for Scotland. He also went a step farther by claiming that *Hibernia* had also once connoted Scotland. Chambers raised the ire of Philip O'Sullivan, who wrote the *Tenebriomastix* to refute his allegations.

**Florence Conry (1560–1629),** archbishop of Tuam, patriot, theologian and founder of the Irish (Franciscan) College of St Anthony at Louvain. His early studies were made on the Continent, in the Netherlands, and in Spain; at Salamanca he joined the Franciscans. In 1588 he was appointed provincial of the order in Ireland and as such sailed with the Spanish Armada. He was sent by Clement VIII to counsel and influence the Irish and their Spanish allies during O'Neill's rebellion. After the disaster of Kinsale (1601) he accompanied Hugh Roe O'Donnell to Spain. He was consecrated Archbishop of Tuam in 1609 at Rome. In 1614, he wrote from Valladolid a vigorous remonstrance to the Catholic members of the Irish parliament for their adhesion to the Bill of

Attainder that seized for the Crown the lands left by the 'flight of the earls'. In 1616 he founded the Franciscan college of St Anthony at Louvain. One of Conry's earliest works was a translation from Spanish into Irish of a cathechism known as 'The Mirror of Christian Life', printed at Louvain in 1626 but probably current in manuscript at an earlier date. In 1618 he presented to the Council of Spain Philip O'Sullivan Beare's 'Relation of Ireland and the number of Irish therein'. He wrote several theological works on grace and salvation, all showing a Jansenist slant. His *Peregrinus Jerichontinus* treats of original sin, the grace of Christ, free will, etc. His other works include *De gratia Christi* (Paris 1646), *De flagellis justorum* (Paris 1644), and *De Augustini sensu circa Beatae. Mariae Virginis conceptionem* (Antwerp 1619).

**Thomas Dempster (1579–1625)**, an inventive Scottish biographer and miscellaneous writer, educated at Cambridge, Paris and Douay, where he published an abusive attack on Queen Elizabeth which excited great indignation among his fellow students and led to a rebellion which had to be suppressed by ecclesiastical authority. The work by which he is now best known is the '*Historia Ecclesiastica Gentis Scotorum*' which was first published at Bologna in 1627, two years after his death. It consists of biographical notices of the writers and memorable historic personages of Scotland from the earliest times to the author's own day. Dempster raised the heckles of his Irish contemporaries because of his attempt to amplify the roll call of Scotland's Catholic saints by hijacking Ireland's heritage of Saints and Scholars.

**Meredith Hanmer DD (1543–1604)**, an English historian educated at Oxford, where he obtained a chaplaincy in 1567 and graduated BA 1568, MA 1572, and DD 1582. He travelled to Ireland c.1591, where he quickly rose in ecclesiastical circles, acquiring many benefices. During his residence in Ireland he wrote his *Chronicle of Ireland,* which drew heavily on Giraldus. It was first published by Sir James Ware in 1633.

**John Hooker (1526?–1601)** was educated at Oxford and travelled widely in Germany and Austria. He travelled to Ireland as a solicitor to Sir Peter Carew, and was elected Burgess for Athenry in 1568. Hooker's chief literary labour was then editing and revision of '*Holinshed's Chronicles*', first published in 1577. This reiterated and added to Girladus' attacks on the Irish.

**Henry of Huntingdon 1080–1155**, an English historian, made Archdeacon of Huntingdon in 1009/10. His interest in History was due to a visit to the abbey of Bec in 1139 where he met the Norman historian Robert de Torigny who brought to his notice the *Historia Britanorum* of Geoffrey of Monmouth. Shortly after, he was himself requested by Alexander, Bishop of Lincoln, to undertake the composition of a history using the writings of Bede as a groundwork. This work, called *Historia Anglorum,* was written in 1129 and first printed by Saville in *Scriptores Post Beda* at London in 1596. He writes as an eye-witness of events surrounding the Norman conquest of Ireland.

**Mauritius Ibernicus (c.1460–1513)**, Mauritius a Portu, or Maurice O'Fihely, who became Archbishop of Tuam, is variously described as a native of Clonfert, Galway, or of Baltimore in Cork. Part of his education was received at Oxford where he joined the Franciscans. Later, he studied at Padua where he obtained a degree of Doctor of Divinity. He was appointed Professor of Philosophy at Padua, studying the works of Duns Scotus, on which he wrote a commentary (published in Venice about 1514). He acted for some time as corrector of proofs to two well-known publishers at Venice, Scot and Locatelli, in the early days, a task usually entrusted to very learned men. O'Fihely was acknowledged one of the most learned men of his time so learned that his contemporaries called him 'Flos Mundi' (the flower of the world). As Archbishop of Tuam he attended the first two sessions of the Fifth Lateran Council (1512). He never took up his see in Tuam as he fell ill en route at Galway, and died there in the Franciscan convent. His edition of Scotus' commentary on Aristotle's Metaphysics was widely read in the renaissance.

**Jacob de Vorraigne (1230–1298)**, a medieval hagiographer. In 1244 he entered the Dominicans and soon became famous for his piety, learning and zeal in the care of souls. He refused the archbishopric of Genoa in 1266, but accepted twenty-six years later in 1292. He is best known for his Legendi Sanctorum. The body of the work which contains 177 chapters (according to others, 182) is divided into five sections, viz. From Advent to Christmas, from christmas to Septuagesima, from Septogesima to Easter, from Easter to Octave of Pentecost and from octave of Pentecost to Advent. His work was extremely influential in the later middle ages.

**Paulus Jovius 1483–1552**, a historian born at Como, Italy. He qualified as a medical doctor at Padua before travelling to Rome where he practiced medicine but also devoted himself to historical studies. Due to papal preferment, he had access to rich sources of information and decided to write a history of Europe beginning with the invasion of Charles VIII of France into Italy and the conquest of Naples. When he completed the first part he read it to the Pope, who was so struck by the elegance of the language and the skill of narration that he conferred a knighthood on Jovius. Pope Adrian VI made him a canon of the cathedral of Como and in 1528 appointed him Bishop of Nocera in 1528. He did not print his work until 1550, under the title of *Historiarum sui temporis libri XLV.* The work appeared in two volumes in Florence (1550) and in Basle (1560). His collected works were published at Basle in 1678.

**Geoffrey Keating DD,** a secular priest, born in Munster, of English lineage. Educated abroad, he returned to Ireland where he wrote his *Foras Feasa ar Éirinn*, or *History of Ireland*, in Irish, drawing on his knowledge of the vernacular sources. His history begins from the earliest records down to the Norman invasion. The history was finished probably around 1633–34.

**St Kilian (c.640–689 ? 9th July),** apostle of Franconia and martyr, probably Irish. There are two accounts of his life but the author is unknown. Then the 'Passio Primus' c. AD 840 Kilian was said to be from Ireland, Scottica Tellus, which at that time could be Ireland. The later Passio says Scotia which is also called Ireland. This edition of the Passio was edited by H. Canisius 'Antiquae Lectionis' Tom IV, li (Ingoldstdat 1603) 642–47. The Passio Secunda edited by Surius, 'De Probatis Sanctorum Historiis IV' (Cologne 1573) 131–5. Serarius edited a text of Surius, 'S. Kiliani Franciae orientalis quae et Franconia dicitur apostoli gesta'(Wurteburg 1598). Separius also adopts this text of Surius for his Opuscula Theologica I (1611) 318–321 (reprint). O'Sullivan takes his extract from the Opuscula.

**Peter Lombard (c.1554–1625),** Archbishop of Armagh, born in Waterford to a respectable and wealthy family of merchants and related to the famous Franciscan, Luke Wadding. Lombard commenced his studies in Louvain in 1572 and was later appointed Professor of Theology there. In 1594 he was made Provost of the Cathedral at Cambrai. When he went to Rome a few years later, Clement VIII thought so highly of his learning and piety that he appointed him Archbishop of Armagh in 1601. He also appointed him a domestic prelate, and thus secured him an income. Henceforth, until his death, Lombard lived at Rome. He was active politically on behalf of the exiled Earls of Tyrone and Tyrconnell. His *De regno Hiberniae Commentarius,* which was published in 1632, though written thirty years previously, presented a vindication of the Catholic community and a celebration of its history.

**John Lynch (1599?–1673?)**, historian, was born in Galway to an ancient Irish family. He was educated by the Jesuits and became a secular priest about 1622. He ran a school and acquired a high reputation for classical learning. Appointed archdeacon of Tuam, he lived secluded from the turmoil of civil strife, in the old castle of Ruaidhri O'Conchobair, last King of Ireland. On the surrender of Galway to Cromwell's forces in 1652 he fled to France. The details of his life in exile are unknown but, as some of his works were printed at St Malo. His main work on Irish history was the famous '*Cambrensis Eversus*', a refutation of the works of Giraldus Cambrensus. He also translated Geoffrey Keating's *History of Ireland* into Latin. He wrote under the pseudonym of Gratianus Lucius.

**Hugh McCaugwell** was successor to Peter Lombard as Archbishop of Armagh. He was born in County Down and became a Franciscan friar. He studied at Salamanca in Spain and afterwards for many years governed the Irish Franciscan College at Louvain where he was Professor of Divinity, as also later at the Convent of Ara Coeli at Rome. He became Definitor General of his order. He wrote *Apologia po Johanne Duns-Scoto Adversus Abr. Bzovium* (Antwerp 1620). This was criticised by a Dominican friar, Nicholas Janssenius, so McCaugwell, under the pseudonym of Hugh McGuinness, one of his pupils wrote in reply: *Apologia Apologiae pro Johanne Duns-Scoto, scriptae advesus Nicholaum Janssen ium ordinis praedicatorum* (Paris 1623).

**Raymond Marlianus**, a commentator on the works of Caesar. He wrote: *C lvlii Caesaris Commentariorum de bello gallico Libri VIII. Civili. Pompeiano, Lib. III. Alexandrino, Lib. 1, Hispaiensi Lib. I. Galliae ac hispaniae: Avarici, Alexiae, Wxelloduni, Massiliae ac Pontis in Rheno Pictura. Locorum insuper, urbium, populorum nomina tum vetera, tum recentiora, copiosissimis indicibus explanata.* This was published at Lyon by Sebastian Gryphius in 1543. The volume concludes with an *Index populorum ac locorum*.

**Pomponius Mela**, a Spaniard from Baetia, was the first Latin geographer. He wrote his *De Situ Orbis* around 43 AD. His writings were mainly a compendium of early Greek writers. He wrote a somewhat fabulous account of Ireland. His work may have been used as a source material by Pliny, the Elder.

**Thomas Messingham (died c.1638)**, a.k.a. Maguinness, an Irish hagiographer, born in the Diocese of Meath. He studied in the Irish College, Paris. He was one of the staff of the Irish College at Paris and was commencing his studies on Irish saints. In 1620 he published Offices of Saints Patrick, Brigid, Columba, and other Irish saints, and the following year was appointed rector of the College, Paris. He was honoured by the Holy See and was raised to the dignity of Prothonotary Apostolic and acted as agent for many Irish bishops. In 1624 he published, at Paris, his most famous work on Irish saints, the *Florilegium Insulae Sanctorum,* containing also an interesting treatise on St Patrick's Purgatory in Lough Derg.

**Sebastian Munster (1489–1552),** a German geographer, mathematician and hebraist who left the Franciscan Order to became a Lutheran c.1529. Shortly afterwards he was appointed court preacher at Heidelberg where he lectured in Hebrew and Old Testimony Exegesis. From 1536 he taught at Basel where he published his great geographical compendium, the *Cosmographia Universalis,* in 1544. His other geographical works are his *Germania Descriptio* of 1530, his *Novus Orbis* of 1532, his *Mappa Europae* 1536, his *Rhaelia* of 1538, and his editions of Solinus, Mela and Ptolemy 1538–1540. His published maps number 142. O'Sullivan's quotations come from the *Cosmographia.*

**William of Newburgh**, born 1135–36, but little is known about his life. He was educated in the Augustinian Priory of Newburgh where he subsequently became a Canon and wrote a chronicle, some sermons, and a commentary on the canticles. His major work is the *Historia Rerum Anglicarum,* written in or just before 1196 and ending in 1198, in which year he probably died. He started his chronicle at the Norman Conquest. The work is of value today, partly because it used the biography of Richard I which is now lost, and partly because of the original material it contains.

**Roderick O'Flaherty (1629–1718),** born in the castle of Moycullen, Co. Galway, the ruins of which are still standing. Under the government that was established by the parliament of England after the Civil War, O'Flaherty was deprived of much of his property. In 1685 he published in London a quarto volume with the following title: *Ogygia seu rerum Hibernicarum Chronologia.* In this work the author treats of the history of Ireland from the earliest times to 1684, with synchronisms and chrono-genealogical catalogues of the Kings of England, Scotland and Ireland to the time of Charles II. His work was the first in which Irish history was placed in a scholar-like way before readers in England and it found its way into many good English libraries of its period.

**Orosius (c.417),** a Christian priest of Spain who accepted from St Augustine the task of showing, in opposition to the adherents of Paganism, that the calamities of his time were neither unusual nor a proof of the anger of the Gods at the spread of the new religion of Christianity. To his survey of human history, from his point of view, he prefixed a geographical essay.

**David Rothe (1573–1650)**, or Donatus O'Ruark, the Catholic Bishop of Ossory. He was chiefly educated at Douay where he graduated in Divinity and returned to Ireland about 1609. In 1616 he published the first part of the *Analecta Sacra,* the second part appeared in 1617. In 1619 Rothe published a third part under the title *De Processo Martyriali,* and the entire work remains as an impeachment of English ecclesiastical policy in Ireland under Elizabeth and James I. He was very active in the Confederation of Kilkenny and John Lynch says that he was the man chiefly instrumental in giving form and order to the Confederacy. He met the Papal Nuncio Rinuccini at the door of St Canice's Cathedral on 11th November 1645. Rothe and Rinuccini did not see eye-to-eye and when Rinuccini left Kilkenny he urged Rothe's suspension. Besides the Analecta Rothe published *Brigida Thaumaturga* (Paris 1620 and Rouen 1621); *Hibernia Resurgens* (Rouen 1621). *De Nominibus Hiberniae Tractatus* and *Elucidationes in Vitam S. Patricii a Joselino Scriptam* are printed in Messingham's *Florilegium Insulae Sanctorum* (Paris 1624).

**St Anthony (1389–1459)**, Archbishop of Florence. He began his career as a zealous promoter of the reforms inaugurated by Bl. John Dominic. In 1414 he was vicar of the Convent of Foligno, then in turn sub-prior and prior of the Copnent of Crotona, and later Prior of the Convents of Rome (Mineva), Naples (Saint Peter Martyr), Gaeta, and Sienna. He was made archbishop of Florence in 1446. He was canonized by Adrian VI on 31st May 1523. His principle work was the *Summa Theologica Moralis, Partibus IV Distincta.*

**Isidore of Seville (560–636)**, born in Cartagena in Spain. Isidore set out to include the Goths in his religious and educational reforms. He presided over the Fourth Council of Seville which begun on 5th December, 633, when he was 73, which commanded all Bishops to establish seminaries in their cathedral cities similar to the one in Seville. He was a prolific writer on sacred and secular subjects. His most famous work is the *Etymologiae,* an encyclopaedia that contained twenty books in all, including: trivium and quadrivium; grammar and rhetoric; dialect; medicine; law; ecclesiastics; God; the church; languages and peoples; etymology; man, beasts and birds; the world and its parts; physical geography; public buildings and road making; stones and metals; agriculture; terminology of war, of jurisprudence and public games; ships houses and cloths; victuals, domestic and agricultural tools. It was a very popular book and was reprinted several times between 1470 and 1529.

**Caius Julius Solinus 235 AD–268 (?)** wrote the *Collectanea Rerum Memorabilium* (Collection of matters of note, published c.230–240). As regards Solinus himself, little or nothing is known. His writings are derived mainly from the 'Natural History' of Pliny and the '*De Situ Orbis*' of Pomponius Mela. Solinus gives a description of Ireland which O'Sullivan uses. Solinus was also called Polyhistor or 'teller of various tales'. He provided the standard source of geographic myth from the fourth to the fourteenth century. His maps were in use right up to the Age of Discovery.

**Edmund Spenser (1552–1599)**, an English poet and administrator in Ireland. In 1578 he became a member of the household of the Earl of Leicester whose nephew was Sir Philip Sidney. Their friendship blossomed and in July 1580, through the influence of these two men, Spenser was appointed Secretary to Lord Grey, then going to Ireland as Lord Deputy. Spenser accompanied Lord Grey on his expedition to Dún an Óir, Kerry, in November 1580 when the Spaniards and Italians who had landed there were surrounded and, though they surrendered and begged for mercy, they were executed and thrown over the cliffs. Spenser fully approved of the massacre. His friend, Walter Raleigh, was one of the leaders of the executioners. In the plantation of Munster he received land as an undertaker at Donerail, Kilcolman Castle, in 1586 when he got 3028 acres, he settled there in 1588. He wrote *The Faerie Queene* while living at Kilcolman. His *A View of the Present State of Ireland,* written in 1596, remained in manuscript form until published by Sir James Ware in 1633. It is one of the most severe arguments for martial law in Ireland, advocating suppression of the native Irish and Catholics by force.

**Richard Stanihurst (1547–1618)** (or Stanyhurst), born in Dublin and educated at Oxford, where he graduated with a Bachelor of Arts degree in 1568. Following this he studied Law at Furnivall's Inn and Lincolns Inn, London. He published his first work, a Latin commentary on Porphyry in 1570. At Oxford he met Edmund Campion who returned to Dublin with Stanihurst and stayed in his house. While staying, Stanihurst helped Campion with his History of Ireland. Stanihurst himself contributed a *Description of Ireland* and *The History of Ireland under Henry VIII* to *Holinshed's Chronicles* (London 1577). In 1581 he left for the Netherlands never to return to Ireland. There he wrote his work '*De Rebus in Hibernia Gestis*' (Antwerp,

1584). The Appendix to this work is based on the topography of Giraldus Cambrensis and repeats the denigration of the Irish. Stanihurst did try to undo some of the worst excesses of Cambrensis in explanatory notes at the end of the chapters. It was against this book that Don Philip O'Sullivan Beare directed his anger in Book V of his Zolomastix. Later in life he is said to have retracted his earlier criticisms of Ireland, and he became a Catholic chaplain to Philip II and subsequently to the Archdukes in the Spanish Netherlands.

**Walafrid Strabo (c.808–899)**. A monk, educated at the Monastery of Reichenau, near Constance. He was appointed by Louis the Pius as tutor to his son, Charles the Bald. He wrote theological historical and poetic works. His chief historical work is a rhymed life of St Gall. Though written 200 years after the saint's death, it is still the primary authority on his life. He wrote a much shorter life of St Othmer, Abbot of St Gall (d.759). His poetical works include a short life of St Blathmac, a high born monk of Iona, murdered by the Danes in the first half of the eighth century. His most famous poem is *The Hortulus,* dedicated to Grimald. It is an account of a little garden he used to tend with his own hands. His book, *De Rebus Ecclisiasticis,* a valuable source of liturgical antiquities, contains a noteworthy passage regarding the Irish custom of repeated prayers and genuflections.

**Laurentius Surius (1522–1578)**, a hagiographer born in the Hanseatic city of Lubeck. It is not clear if his parents were Catholic or Lutheran. He studied at the universities of Frankfort on the Oder and Cologne. He devoted himself chiefly to the domains of church history and hagiography and wrote a large number of works on these subjects. Surius includes a Life of St Romuld written by Theororicus under 1 July.

**Cornelius Tacitus (c.54/56–117?)**, a Roman historian and a friend of Pliny the Younger. He was writing his 'histories' between 104 and 109. He was writing the 'annals' in 117, but death prevented him from writing about the happy reigns of Nerva and Trajan. His father-in-law was Agricola who was Governor of Britain from 78–86 AD. Agricola looked at Ireland and estimated that all he needed to conquer it was one legion plus auxillaries. Tacitus gives a brief, and in parts obscure but interesting, description of Ireland.

**Theodoricus (c. 1100–1150)**, a Platonist philosopher born in France. He was head of the School of Chartres in 1121. He later seems to have gone to Paris when John of Salisbury was one of his pupils. In 1141 he was again teaching in Chartres. He wrote a book on the seven liberal arts, entitled *Hepta-teuchon,* a treatise on the creation *De Sex Dierum Operibus* and a commentary on *De Inventione et Rhetorica ad Herennium.* He wrote a life of St Romuld or Rombaut honoured as the Apostle of Malinesy Mechlin, Belgium. Romuld, according to tradition, was a hermit who was murdered. Theodoricus states that Romuld was a bishop of Dublin. Romuld died c.775 AD.

**Vincent of Beauvais**, a priest and encyclopaedist. Little is known of his personal history. The years of his birth and death are uncertain, the dates frequently assigned vary from 1190–1264 respectively. It is thought that Vincent joined the Dominicans in Paris shortly after 1218; with the exception of visits to Louis IX at Royaumont he spent all his religious life in the monastery at Beauvais. A man of industry, Vincent undertook a systematic and comprehensive treatment of all branches of human knoweldge. In the preparation of this colossal work he was helped by the purchase of books by his patron, Louis IX. The general title of Vincent's work is *Speculum Majus.* It contains eighty books divided into 9,885 chapters, figures which give some idea of the magnitude of the work accomplished by the Dominican Friary in the first half of the thirteenth century.

**John Wadding**, a secular priest of Wexford, John Wadding wrote a treatise entitled *Historia Ecclesiastica Hiberniae* against Thomas Dempster, a Scotsman but an ecclesiastic of Bologna, who provoked many of the Irish to oppose him on account of a book he wrote entitled *Nomenclatura Scriptorum Scotorum,* wherein he inserted a great number of Irish saints and writers and passed them off as his countrymen.

**Luke Wadding (1588–1657)**, an Irish Franciscan born in Waterford. In 1604 he went to study in Lisbon and at Coimbra. He resolved to join the Franciscans and spent his novitiate at Matozinhos. He was ordained a priest in 1613. In 1617 he migrated to Salamanca where he became President of the Irish College. He went to Rome in 1618 as Chaplain to the Spanish Ambassador and there resided until his death. He collected funds and on 24th June 1625 founded the College of St Isidore for Irish students in Rome. He gave the College a library of five thousand printed books and eight hundred manuscripts and thirty resident students soon came along. Wadding was Rector for fifteen years. He was an enthusiastic supporter of the Irish Catholics in the war of 1641 and his college became the strongest advocate of the Irish cause in Rome. Wadding sent officers and arms to Ireland and induced Innocent X to send as

Papal Nuncio Giovanni Battista Rinuccini to the Confederation of Kilkenny. The Confederate Catholics petitioned Urban VIII to make Wadding a Cardinal but the Rector of the Irish College found means to intercept it and it remains in the archives of the College. He published in all thirty-six volumes; fourteen at Rome, twenty-one at Lyon and one at Antwerp. A great interest was the Immaculate Conception of the Virgin Mary.

**Hugh Ward (d.1635)**, born in County Donegal in Ulster but educated partly at Salamanca and partly at Paris, and afterwards was made a First Lecturer and then Guardian of the Irish College of Louvain; having before been admitted to the Order of Franciscan Friars at Salamanca in the Year 1616. He undertook to write a general history of the lives of the saints of Ireland, for which he employed Michael O'Clery and sent him from Louvain to Ireland in search of manuscripts and to gather materials for the work. He died before it was finished, but his papers proved of singular use to John Colgan who afterwards applied himself to the same subject. About a year after his death was published a treatise of his, entitled *Dissertatio Historica de S. Rumoldi Patria. Louvain 1616, 470.* He wrote several pieces as preliminary to his larger work (which later became *The Annals of the Four Masters*).

**Stephen White (1575–1647)**, a Jesuit, born in Clonmel. He was educated at the Irish seminary at Salamanca where he was a reader in Philosophy. He joined the Jesuits in 1596 and in 1606 he became Professor of Scholastic Theology at Ingoldstadt and returned to Spain in 1609 but did not live there long. He is chiefly remembered for his labours among Irish manuscripts preserved in German monasteries. He corresponded in a friendly way with the Protestant Archibishop Ussher who acknowledges his courtesy and testifies to his immense knowledge, not only of Irish antiquities but of those of other nations. He returned to Ireland in later life and died there. His best known work is the *Apologia pro Hibernia adversus Cambri Calumnias*. It is believed to have been written as early as 1615 and was supposed to have been lost. John Lynch used an imperfect copy for his '*Cambrensis Eversus*'. A manuscript copy of White's has recently been discovered at the Bibliothèque Municipal in Poitiers entitled *Apologia pro Innocentibus Ibernis Olim Temere Traductus per Richardum Stanihurstum*. White criticises Stanihurst severely.

# Index

Compiled by Julitta Clancy

*Note:* page references in *italics* denote translator's introduction and footnotes

284